WIRELESS AT WAR

Developments in Military and Clandestine Radio

1895-2012

BF trench set in operation on the Western Front (RAS)

WIRELESS AT WAR

Developments in Military and Clandestine Radio
1895–2012

Peter R Jensen

ROSENBERG

First published in Australia in 2013
by Rosenberg Publishing Pty Ltd
PO Box 6125, Dural Delivery Centre NSW 2158
Phone: 61 2 9654 1502 Fax: 61 2 9654 1338
Email: rosenbergpub@smartchat.net.au
Web: www.rosenbergpub.com.au

© Copyright Peter R Jensen 2013

All rights reserved. No part of this publication may be reproduced, stored in a retrieval system, or transmitted, in any form or by any means, electronic, mechanical, photocopying, recording or otherwise, without the prior permission of the publisher in writing.

Every effort has been made to trace owners of copyright of material included in this book, but advice of any omissions would be appreciated.

National Library of Australia Cataloguing-in-Publication entry

Author: Jensen, Peter R., author.

Title: Wireless at war / Peter R. Jensen.

ISBN: 9781922013477 (paperback)

Notes: Includes bibliographical references and index.

Subjects: Telegraph, Wireless--History.

Military telecommunication--History.

Dewey Number: 384.52

Cover: PRC-F1 SSB synthesised HF radio by AWA (RAS)

Printed in China by Prolong Press Limited

Contents

Acknowledgments   7
Key to illustration sources   9
Introduction   10

Part 1: 1895-1920   12
1   Wireless Beginnings: a new means of communication   12
2   War Wireless Before 1914   29
3   The First World War   50
4   Technological Change: from spark to valve   82
Project 1: Wilson transmitter replica   95

Part 2: 1921-1950   107
5   The Interwar Years   107
6   Interwar Military Communications   115
7   World War Two   124
8   Radio Communications in Australia's war zones   133
9   Clandestine Communications   140
10   Technological Change: valves, miniaturisation and circuitry   160
Project 2: Paraset replicas   176

Part 3: 1951-1970   189
11   After World War Two   189
12   Computers, *Sputnik* and ARPANET   204
13   The Vietnam War   212
14   The Solid-state Revolution   222
15   Technological Change: from valves to transistors and integrated circuits   229
Project 3: Solid-state double-sideband transceiver   242

Part 4: 1970-2012   250

16   Towards a New Century: changing warfare   250
17   Military Communications Requirements   255
18   Technological Change: Digital Development, Encryption, Jamming and SDR   262
19   Battlefield Communications: 1976-2012   272
20   Contemporary Military Communications   280
21   Future Directions   286

Postscript   291
Bibliography   293
Glossary of Abbreviations   304
Appendix A: Tactical Radio Database   08
Appendix B: Large Schematic Diagrams   318
Index   347

Acknowledgments

Many persons and organisations provided assistance and material in the creation of this work, a process that has continued over a number of years. Keeping track over time of all sources of material has proved difficult, and it is very possible that some have been overlooked, for which I offer my sincere apologies.

Firstly, acknowledgement of the assistance received at the Royal Signals Museum at Blandford Forum in the Great Britain and the curator at an earlier time, Major Roger Pickard, is particularly appropriate. Because the Australian Army was so intimately related to the British Army until the 1950s and 1960s, much of the early wireless apparatus was of British origin and often locally modified for Australian and tropical conditions. This relationship only started to lessen with Australia's collaboration with the United States in the wars in Korea and Vietnam.

In addition to the Royal Signals Museum, access to the Amberley Chalk Pits Museum via its curator, Mr David Rudram, provided illustrations of interest, as did the Imperial War Museum.

Access to the Royal Australian Signals Museum at Watsonia, north of Melbourne, has been immensely valuable in providing information relating to more recent warfare in which Australia has been involved. In particular the assistance of the curator and manager, Major Jim Gordon, in allowing access to equipment, and in scanning slides and other material in the museum, is gratefully acknowledged.

In the late 1980s, access to the collection of World War One wireless apparatus owned by Mr Bill Journeaux of Poole in Dorset allowed images of extremely rare items be made. More recently, the remarkable private collection of clandestine radios of Mr John Elgar-Whinney in Kent has helped to consolidate the range of apparatus available during and after the Second World War.

Access to the Royal Tank Museum at Bovington made possible the creation of a graphical database of tanks from all periods and various national sources. This was a valuable resource with which to expand the text in relation to armoured mobility and communications.

At the Duxford branch of the Imperial War Museum, the Duxford Radio Society's Vice-President, Mr Denis Willis, proved a highly informative and helpful guide to the collection of valuable radio exhibits.

In Canberra, the staff of the Australian War Memorial provided access to particular exhibits, including the De Mole tank, which contributed significantly to the development of the local element of this history.

The support of members of the Historic Radio Society of Australia is also acknowledged and, in particular, the assistance of Mr Ian O'Toole who operates an admirable military communications museum at Kurrajong in New South Wales. Mr Lou Albert provided rare components that made possible the successful completion of two of the projects, the Wilson Transmitter and the Paraset. Another critical element came through the efforts of Mr Ray Robinson, who undertook a meticulous reading and editing of the draft manuscript.

Last but by no means least has been the support of my wife Helen, who has cast a benign eye over the antics of her obsessive husband in his journeys to examine places of historic radio interest both near and far in the task of unravelling the technological changes that have occurred over the last 100 years.

And finally, to provide some light relief from the parade of historical and technical matters, instructions for the construction of practical examples from successive eras of radio development have been included at the ends of Parts 1, 2 and 3. Traditional historians may find this an unnecessary intrusion into the story of military radio, but for many others it may prove a desirable interlude. Moreover, in terms of the evolution of the technology of military radio communications, undertaking these projects may be more revealing than the words and illustrations of the main text.

Key to Illustration Sources
The following organisations and individuals are acknowledged as the source of the photographs contained in this book and are, where appropriate, thanked for allowing the author to include their material in this publication.

AMP	Australian military publication
AWA	Amalgamated Wireless (Australasia) Ltd
BPM	Bletchley Park Museum, UK
BUR	K Burke, With Horse and Morse in Mesopotamia
HAR	Harris Technology, USA
ILN	Illustrated London News
INT	Internet source (copyright holder unknown)
KMM	Kurrajong Military Museum Archives, NSW, Australia
LM	Louis Meulstee, UK
MAR	Marconi Archives, Chelmsford, UK
NOR	Norsk Hydro Museum, Norway
PRJ	Peter Jensen photographic archives
PRO	Denis Hare, BEM, Pronto in South Vietnam 1962-1972
RAS	Royal Australian Signals Museum
RAY	Raytheon Corporation, USA
SEL	SELEX Marconi Corporation, UK
SMB	Signals Museum Blandford Forum, UK
UNK	Unknown source
USM	United States military publication
WWW	Wireless World

Sources of copies made by the author
Permission has been sought to use this material and where received this is expressly acknowledged with appreciation.

AOB	Sampson Low Marston
BAW	Hutchinson Publishers
BBC	British Broadcasting Commission
BOS	Thomas Nelson & Sons
DKI	Lovat Dickson
GAL	Heinemann
MMS	George G Harrap
MWT	Naval Institute Press
PRR	Funk & Wagnalls
SAW	Her Majesty's Stationery Office
SWA	CR Leutz
WIT	Oxford University Press
25Y	Odham's Press
40B	Hutchinson Publishers

Introduction

Since 1895, when the history of wireless communication began, some of the most inspired and significant technological advances have been spurred by warfare. From this painful circumstance have come highly desirable and beneficial technical results and changes in the longer term.

Following the demonstration by Heinrich Hertz in 1885 of the existence of electromagnetic radiation, this new field of science was at first investigated by a relatively small group of experimenters, supported by the early telecommunications industrial entities such as the Marconi Company in Great Britain, and Telefunken in Germany.

After 1900, as the potential advantages of wireless technology to the battlefield and for marine warfare became apparent, interested industrial organisations multiplied, keeping pace with the growth of the armies and navies that would soon apply their products to devastating effect.

How twentieth-century warfare helped to speed the development of radio and electronic communications is worthy of recall. This book's intended audience includes the current generation of readers who have never been without access to the telephone or the microcomputer or the Internet.

The forces that have helped to create the convergence of electrical, electronic and electromagnetic means of communications have been discussed in other places by this author and many others. Warfare, however, can be seen as having had a more profound impact on this process over the past 100 years than almost any other driving force. Describing how this impact has led to improved systems of communications and a progressive diminution in size and weight of apparatus is one of the main objectives of this volume.

In the distant future historians may assess the twentieth century as an exceptionally turbulent and violent period that involved warfare throughout almost the whole of its span. More optimistically, it may also be seen to have been a century of extraordinary development in telecommunications and computing.

In presenting this examination of the development of military wireless (or radio as it is referred to in general), it will be seen that initially Great Britain is cited as a major source of Australian inspiration; only later does America appear in the context of military radio used in Australia. This reflects the developments in Australia's international relationships. Until after the Second World War, Australia remained very much a colonial element of the diminishing British Empire. When the United Kingdom joined the European Common Market, this constituted the 'cutting of the apron strings' that Australia had been surprisingly reluctant to initiate on its own behalf.

Since the 1950s, geographical realities and the emergence of bitterly fought wars in Southeast Asia, and Vietnam in particular, have furthered an alliance with the United States that was initially forged in the Second World War. In more recent times it has led to Australian involvement with the United States in its pursuit of change in Afghanistan and the destruction of al-Qaeda.

Part 1: 1895–1919

1 Wireless Beginnings: A New Means of Communication

The discovery of X-rays by German physicist Wilhelm Röntgen in 1895 and, almost simultaneously, the experimentation of Guglielmi Marconi at the Villa Griffone at Pontecchio in Italy which led to wireless telegraphy, may be seen as events that were an entirely fitting culmination to a century and more of intense scientific and technological development—which was about to give way to a century of a very different character. The twentieth century may be better remembered for its pervasive wars than its many notable scientific and technological advances. The second decade of the century saw the terrible violence of the First World War and the horrors of trench warfare. Just twenty years later, this dreadful example, the 'war to end all wars', did not stop humanity from once again resorting to violence with the outbreak of the Second World War.

From the summer of 1895, when the success of the first experimental wireless transmission was announced by a gunshot at Villa Griffone, it is possible to chart the part that wireless would play in supporting the violence of the twentieth century and warfare in particular.

Marconi's initial experiments converted the laboratory experiments of Oliver Lodge and Heinrich Hertz into a functional communications system that did not require kilometre upon kilometre of cable and telegraph wire, instead relying on the intangible forces of radio frequency energy to convey the dots and dashes of the Morse code. The existence of the radio frequency field and energy propagation had been anticipated many years earlier in the remarkable and fertile imagination of one of the Victorian era's most renowned experimenters and scientists, Michael Faraday.

Finding the Waves

In the early part of the nineteenth century, Faraday undertook a series of experiments which led to a fundamental understanding of the relationship of electricity and magnetism. From these experiments came the electric motor and the dynamo, which allowed electricity to be used to drive machinery and also to be generated by the application of rotational energy. Later, this work led Faraday to speculate on the physical relationship between magnetism and electricity. An 1846 paper entitled 'Thoughts on Ray Vibrations' reveals that his thoughts had turned to the question of light and its propagation. Although the ideas expounded in that paper were significantly at odds with contemporary scientific attitudes and knowledge of the subject, at least in one place they were to find fertile ground in which to take root—in the mind of the young and remarkable Scottish physicist, James Clerk Maxwell.

Maxwell's dynamical theory of the relationship between electricity and magnetism, now referred to as electromagnetic radiation, can be seen as providing the intellectual foundation for radio communications, or wireless telegraphy as it was better known in the early years. In a paper presented to the Royal Society in 1864, Maxwell provided a series of equations to describe the interaction between electricity and magnetism. These equations allowed him to predict the existence of waves created by the oscillation of electric and magnetic fields, which would travel at a speed equivalent to that of light at approximately 300 000 000 metres per second. From this prediction came the realisation that electromagnetic radiation involved the same physical phenomenon as light—the transmission of waves of energy through a hypothetical medium referred to in Victorian times as the 'ether'. More recent methods of measurement have led to the knowledge that light travels at a fixed velocity of 299 792 458 metres per second as determined by Albert Michelson in 1879.

Some years after the publication of Maxwell's theoretical work, other eminent scientists, notable among them being Hertz and Lodge, undertook a number of experiments and demonstrations that established the fundamental characteristics of electromagnetic radiation. Their work confirmed both the physical reality of electromagnetic waves and their fundamental similarity to light radiation. It was this work, in particular that of Hertz, that was to provide a stepping stone to telecommunications without wires, 'wireless', as developed by Marconi and his contemporaries.

At a later stage too, the failure of scientists to demonstrate the existence of the ether led directly to the theory of relativity as propounded by Albert

Einstein in 1905, and his then science-challenging assertion relating to the constant velocity of light under all circumstances.

Marconi and Wireless

How Guglielmo Marconi, a mere youth, became aware of the existence of electromagnetic radiation and then produced a system of wireless communication in the eighteen months following the premature death of Hertz in 1894 has been told by a number of writers, including the present author. To recap briefly, following the rebuttal of his experimental work by the Italian postal authorities, Marconi came to England with his British mother, Annie Marconi, where he was listened to with considerable interest by the chief electrical engineer of the Post Office, William Preece.

Apparatus destroyed by over-enthusiastic officers of Her Majesty's Customs on Marconi's arrival in England (they suspected it was related to bomb-making) was repaired, and demonstrations organised. At one of the earliest, held on Salisbury Plain on 2 September 1896, two particularly interested observers were a future admiral of the Royal Navy, Captain Henry Jackson, and a military officer, Major Carr.

Also in attendance that day was the German scientist Adolf Slaby, from the Berlin Technical University, who in later years was involved in the creation of the industrial enterprise now known as Telefunken. This derived from an initial business partnership between Slaby and Count Georg von Arco in the business known as Slaby-Arco, and the Siemens & Halske business in which Professor Braun was involved. Telefunken was to become a formidable rival of the Marconi Company. Telefunken survives to the present in various guises and locations while, tragically, the residual elements of the original Marconi Company have imploded. In 2006, what remained was absorbed into the Ericsson telecommunications giant.

It seems highly probable that the representatives of the British Army and Navy were influential in the adoption of wireless apparatus by both services. Before he became aware of Marconi's apparatus, Captain Jackson had developed a form of wireless telecommunications apparatus for use at sea as a response to the problem of controlling the movements of the newly developed, high-speed motor torpedo boats. Jackson's system was considerably less effective than Marconi's, thus his interest is not hard to understand. The potential advantages for military operations of communications apparatus that did not require a wired connection must have been just as obvious to Major Carr.

Wireless Beginnings: A New Means of Communication

Marconi demonstrating his apparatus on Salisbury Plain, 1896 (MAR)

The Marconi wireless communications system was patented in 1896 in a very comprehensive, diagrammatically illustrated statement of its capabilities. In Patent Number 12039 of 1896, the system was described as involving the 'effective telegraphic transmission and intelligible reception of signals produced by artificially-formed Hertz oscillations'.

Very soon the system's commercial possibilities were perceived, and with the involvement of Marconi's cousin, Henry Jameson-Davies, and members of his mother's extended British family and their business connections in the whisky distilling business, the Marconi Company was formed. (It was founded in 1897 as The Wireless Telegraph & Signal Company, and over the years underwent many changes of name. For simplicity it is referred to here as the Marconi Company.) For much of the next few years, Marconi was engaged in a continuous round of newsworthy activities in which the new wireless telegraph was the main attraction. A number of these activities involved communications paths over water or with marine craft where a conventional cable connection was impossible. Perhaps the most notable early activity was the presentation of wireless communications to the public in a lecture at the Toynbee Hall, London, on 12 December 1896. This same date in 1901 would become even more notable when a signal made with the letter 'S' of the Morse code was sent across the Atlantic from Cornwall to Newfoundland using a spark generated by the Marconi transmitting apparatus.

The Toynbee Hall lecture was given by William Preece with Marconi's assistance, carrying his now famous 'Black Box' receiver around the hall. As Preece pressed a Morse key at the dais, a large induction coil to which it was connected discharged a stream of sparks forming the letters of the Morse code. This action was responded to by a bell on the Black Box and, as reported by the newspaper representative present, caused a considerable sensation. The reporter submitted copy that resulted in a headline referring to Marconi as the 'inventor of wireless', a title he had not claimed. As Preece had noted during the course of the lecture, no new devices had been presented as part of the Marconi communications system. What Marconi had achieved was the development and association of existing devices (created by Lodge, Branly and others) to produce a communications system that was fully functional, not just theoretical. It was this which Preece was able to present to the public as a novel and highly significant development.

Marconi with the Black Box, 1896 (MAR)

Using the analogy of Christopher Columbus's egg, Preece said that Marconi had shown how the Hertzian wave 'egg' could be used to achieve distant communications which, up to that time, had been discussed but not demonstrated. Just as Columbus had showed a way to make a boiled egg stand on its end by breaking the shell, Preece explained that what Marconi had accomplished had become 'obvious'—but until his experiments had been completed, it was not. Communicating without wires was something that no one else had yet achieved, despite experiments that had shown the existence of Hertzian waves capable of conveying intelligence. Preece was probably referencing the work of Oliver Lodge in particular, and to a lesser extent that of Serbian-American electrical engineer Nikola Tesla, the Russian physicist Alexander Stepanovich Popov, the French scientist Edouard Branly and others.

Despite dismay and publicly expressed outrage at the newspaper's characterisation of Marconi as the inventor of wireless, later in 1897 Lodge did have the good grace to concede in relation to his own work that 'no attempt was made to apply any but the feeblest power so as to test how far the disturbance could really be detected' (Simons 1996).

Lodge also admitted his own stupidity in not seeing the potential of Hertzian waves for communication purposes. One can assume he was referring to an experiment he carried out in June 1894 in which Morse signals were sent over a distance of 150 yards. This had occurred during a lecture to the Royal Institution using a coherer detector as originally developed by Branly. Within a year of Lodge's experiment, Marconi had been able to send a Morse code signal over a distance of approximately 1.5 kilometres (just under a mile). He was assisted by his elder brother Alfonso, who indicated the receipt of the signal by a gunshot. The success of the experiment led Marconi to expose his work to public scrutiny.

From his very first appearance in Great Britain, Marconi showed a considerable talent for self-publicity and in using his apparatus to undertake tasks of communication that previously had not been possible. Marconi was able to secure the patronage of royalty and, during the Cowes Regatta of 1898, installed his system both on the Royal Yacht *Osborne* and at Osborne House on the Isle of Wight. He was able to transmit and receive some 150 messages between Queen Victoria and her son, the Prince of Wales, a service that received both royal appreciation and exposure in the press.

Far more significant to the future military use of wireless communications was the decision of the Royal Navy to employ the new system during manoeuvres in 1899, using it to link ships over a maximum distance of 95 miles (150 km). As was apparent to Captain Jackson, in command of one of the vessels, this involved sending wireless signals beyond the curve of the earth. During the exercise HMS *Europa* had maintained communication with HMS *Alexandra* at a location far beyond the horizon. This contact was assisted by HMS *Juno*, which lay between the other two ships and acted as a wireless relay station. Jackson's observation was to be highly relevant to the trans-Atlantic experiment of 1901, when the Morse letter 'S' was transmitted from Cornwall to Newfoundland.

In the final years of the nineteenth century, Marconi's time was largely taken up by a continuous round of publicity-seeking activities, all intended to ensure the success of wireless as an alternative to earth-bound telegraphic systems. Communication with locations inaccessible with conventional cables or wires became the principal concern of his new commercial organisation.

Signals were sent across the Bristol Channel from Penarth in Wales to Brean Down in England. This experiment had initially involved the two islands in the Bristol Channel, Flatholm and Steepholm, as the sites of receiving stations, but vertical extension of the antenna system ultimately allowed the full width of the waterway to be crossed and Morse-coded messages sent from Wales to be received in England.

The next significant transmission over water was between England and France, across the English Channel. This experiment was conducted between the South Foreland lighthouse in Kent and a temporary antenna erected on the foreshore of the small seaside town of Wimereux, just north of Boulogne.

While these experiments were important, access to marine vessels was an even more obvious and relevant target for telecommunications without wires, and from 1898 to the turn of the century, marine wireless stations became one of Marconi's particular concerns.

During this period, Marconi crossed the Atlantic to demonstrate wireless to American yachtsmen and to promote the new technology as a prelude to opening up commercial activities in the United States. As this visit coincided with his experimental work into methods of 'tuning', which had to be kept secret, requests for technical information from potential American users were refused, and Marconi returned to Great Britain without having established a commercial opening.

Naval Manoeuvres

In the longer term and, as compared with these somewhat trivial activities, far more important was the introduction of wireless into the Royal Navy. Captain Jackson, as an early investigator and experimenter in wireless propagation, had been present at the demonstration on Salisbury Plain in 1896. As a modern and keen exponent of the scientific method as it could be applied to the Navy, Jackson had strongly supported the wireless experiments that were undertaken in 1899. Despite some fraught negotiations relating to the royalties to be charged for use of the Marconi system, in that year wireless apparatus had been installed in three ships of the Royal Navy. The success of wireless in maintaining communications over a distance of 150 kilometres elicited a strong recommendation from the fleet commander, Vice-Admiral Domville, that the Navy adopt the system.

As the accompanying illustration demonstrates, a marine wireless station of this period was simplicity itself. One pole of a Ruhmkorff coil was connected to an aerial running up into the rigging of the vessel, and the other pole of the coil was connected to the sea-water, acting as an earth. This constituted the primary transmitting circuit which in the first instance was completely untuned other than through the length of the aerial wire. Later the primary transmitting circuit would include a device that at the time was called a 'jigger'—this enabled a match of the impedances of transmitter and aerial circuits to be achieved. In the jigger a pair of inductances, coils which could be moved relative to each other, were included; this allowed inductive matching of the local spark circuit to the aerial impedance to be obtained.

Importantly, it was apparent that matching of

Early untuned ship's transmitter (MAR)

the inductances allowed significantly increased power output from the marine wireless station as well as increasing the effective range of the signal transmitted. The coherer referred to earlier remained the means of converting radio frequency energy into sound for a few years, but after 1902 it would be replaced by a more robust and less temperamental device, the magnetic detector.

Mobility and Wireless

Prior to Marconi's Salisbury Plain demonstration, whether on the battlefield or during a naval action, keeping in touch while mobile was a difficult and frequently impossible task. On land, message transmission might well involve a messenger on horseback, a mechanical semaphore or a heliograph. After the inventive work of Samuel Morse, it could also involve a telegraph cable and the passing of Morse code messages with a hand key. At sea, however, message handling was only possible within line of sight using signalling lights or semaphore flags, and the range at which such systems could operate was extremely limited. Once a ship was over the horizon, communications became impossible. With the coming of wireless, that situation changed dramatically. For communications between mobile operators, wireless offered a system unmatched by any existing method.

Despite the obvious advantages of mobility, it was the Navy rather than the Army that soon adopted wireless. As later problems in South Africa were to reveal, the Navy had a major advantage in operating the new system: access to the good earth provided by sea-water. This meant there was far less difficulty in adapting wireless to the operational needs of the Navy than to the Army, which moved over terrain of varying electrical character, as had occurred in the Boer War Campaign in 1899 to 1902.

As described in Chapter 2, the limitations of wireless technology were to provide a significant barrier to its effective adoption by the military before the start of the First World War, but as conventional telegraph and telephone wires were torn up by shellfire on the Western Front, wireless became an indispensable battlefield communications aid. The Army's adoption of this system of communication was then inevitable and rapid.

Early Wireless Technology

The receiving apparatus initially developed by Marconi, the famous Black Box, was to modern eyes remarkably primitive. In many ways, the surprise was that it worked at all. In particular, the device that was used to detect the radio frequency energy, Hertzian waves, operated on a physical principle that even today remains somewhat obscure. This was the coherer, whose name immediately suggested its mode of operation. Simply explained, it was a radio frequency-sensitive switch in which metallic particles set between two conductors clung together in the presence of radio waves, thereby creating a short circuit. Once the metal particles had cohered, a sharp tap was necessary to bring them back to the open circuit state.

The coherer formed a part of a circuit which included an electric bell and a battery. The tapper of the bell was set close to the shaft of the glass coherer so that when the bell rang it also lightly tapped the coherer, whereupon the metal particles de-cohered. This mechanical feedback circuit was able to convert radio frequency energy to sound, but only relatively slowly, so that Morse code could be sent at approximately ten words per minute. This was less than half the conventional speed of message handling by wired telegraph. Clearly higher speed was required. The next generation of wireless apparatus saw more stable and more sensitive means of detecting radio frequency energy developed.

In this period the wireless transmitter remained essentially that employed by Hertz in his experiments in the 1880s—the Ruhmkorff (induction) coil. The insertion of a Morse key between the induction coil

Inside the Black Box coherer receiver (MAR)

and a battery allowed Morse code to be generated and transmitted as radio frequency energy generated by the spark between electrodes connected to the induction coil.

One of Marconi's many contributions to the development of telecommunications was the discovery that connecting one electrode of the Ruhmkorff coil to the earth and elevating the other with a wire (in what is now called an antenna or aerial) increased the range over which signals could be sent to a quite dramatic extent. Another was that increasing power in generating the spark had the effect of increasing the distance over which radio waves could be sent, something that Lodge had singularly failed to appreciate. These two discoveries were for some time to obscure the value of short wave lengths, which later became the basis of worldwide telecommunication.

Wireless Development: 1895–1902

The simplicity of early wireless apparatus is readily understood by recourse to schematic diagrams. In an era when it is possible to almost instantaneously convey face-to-face video and audio information from one end of the globe to the other, it must be very hard to believe that such means could be employed to convey intelligence in the form of the Morse code.

The original Marconi apparatus consisted of an induction coil-based spark transmitter and a coherer-based receiver which was able to receive Morse code at about ten words per minute. The receiver employed a system of electro-mechanical feedback to loosen the particles of metal in the coherer once a conducting path had been established by the imposition of a pulse of radio frequency energy.

The major problem with this first-generation communications system was the absence of any method of containing the radio frequency energy produced by the spark gap, which meant that the operation of any other station inevitably interfered with the operation of the first station. The solution to

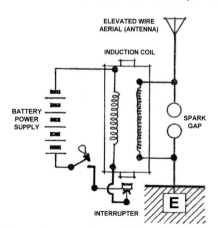
Marconi transmitter, 1896 (PRJ)

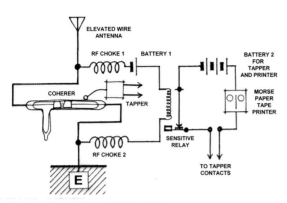

Marconi receiver, 1896 (PRJ)

this problem was called 'syntony' (now referred to as tuning), and was the subject of intense experimentation during the latter years of the nineteenth century. Marconi's initial efforts to tune the broad band of radio frequency energy resulted in the insertion into the transmitter of the crude radio-frequency transformer (inductive coupling) referred to as a 'jigger'. As the accompanying schematics show, this was inserted between the spark balls and the aerial, and provided a degree of coupling which allowed the aerial to resonate at something closer to a single frequency. Nonetheless, the transmitter still produced a signal that later operators would describe as being as 'broad as a barn door'.

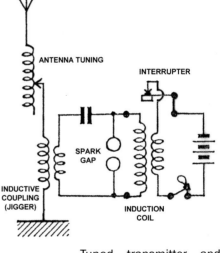

Tuned transmitter and receiver, 1902 (PRJ)

By 1902 the jigger device had been incorporated into the coherer receiver, and as the schematic on the previous page shows these changes in the primitive system did provide a degree of tuning. However, this improvement could only achieve a modest improvement while ever a spark was used to provide the radio frequency energy. Though tuning produced a degree of concentration of radio frequency energy in a selected part of the spectrum, while ever a spark was used to provide it the radio frequency energy thus generated remained, in effect, a diffuse and broadly distributed band of electrical noise.

By 1902, only six years after Marconi had developed the first practical

Marconi coherer of 1902 (MAR)

wireless communications apparatus, the critical early detecting device, the coherer, had become a well-finished industrial product. By then, following pressure from his commercial advisers, it was being manufactured by a newly created industrial concern that bore Marconi's name.

A typical Marconi coherer of the period is illustrated here. The glass bulb of the coherer is positioned horizontally between two supporting columns. Below the glass tube is the brass tapper element attached to the activating solenoids, which are set on a sliding bed with a screwed threaded rod and adjusting knob. This same configuration is also seen in the complete receiving device, the Marconi coherer receiver.

In the coherer receiver, the coherer unit was set on a separate baseboard, together with a box of dry batteries, the relay required to activate the tapper mechanism, and an external bell or paper-tape inker unit (not shown here). In front of the coherer is a box containing capacitors and immediately adjacent to this are the radio frequency chokes. In both land-based and ship-borne installations, this receiver was set up in a metal box with a hinged lid that could be closed to exclude stray radio frequency energy from external apparatus, including the electro-mechanical make and break of electric bells and Morse code inking machines where sparking at contacts occur. A number of these metal boxed installations are still in existence.

A typical marine wireless installation appears in the illustration overleaf,

Marconi coherer receiver of 1902 (MAR)

with the coherer receiver units visible within their metal screening boxes. Between the coherer receivers and the wireless operator and Morse key is a paper-tape inker to record the Morse signals and, behind it, a bank of six Leyden jar condensers (capacitors). Above this array of items, fixed to the wireless cabin wall is a jigger (matching inductance). This device is connected between the high-voltage output from the Leyden jars and the aerial cable, which is taken up to the wire hung between the ship's masts.

Behind the operator's arm and the Morse key is a Ruhmkorff coil with an enclosed spark gap. This tubular device is set between two Ebonite pillars and connected at one end to the Leyden jars and at the other to the metalwork of the ship's hull, providing the necessary earth connection to the external sea-water. The power supply control panel and battery-charging controls are fixed to the cabin wall behind the operator's head.

With this remarkably simple apparatus, and taking advan-tage of an elevated aerial using the mast as its support, a substantial range of communications was possible. Contacts were conventionally possible over 70 to 100 kilometres during the day, and at night frequently over considerably longer distances.

Shipboard wireless installation, 1902 (MAR)

Across the Atlantic

Marconi believed that, contrary to popular scientific opinion, it would be possible to send a radio frequency signal around the curvature of the earth's surface from England to the United States. In 1901, with the assistance of a new adviser to the Marconi Company, Professor J Ambrose Fleming, this remarkable experiment was undertaken. On 12 December 1901, the Morse letter 'S' was heard at a temporary wireless station established close to the Cabot Tower at St Johns, Newfoundland. It had been sent from a site at Poldhu in Cornwall, at the westernmost tip of Great Britain, and generated by a 12 000 watt spark transmitter developed with the assistance of Professor Fleming.

Given the prevailing scientific climate, it is hardly surprising that the news of this success was greeted by some with a degree of scepticism that has persisted even to the present. Disbelief derived from the assumption that radio frequency energy could only be propagated in a straight line, as is the case with light, which in 1901 was the more familiar manifestation of such energy.

The means by which the Poldhu signal was able to bridge the Atlantic and travel around the curvature of the Earth has been examined by a number of persons over the years. To this author, the most useful analysis was presented at an Institution of Electrical Engineers con-ference in London in 1995, to commemorate the first 100 years of radio communications. In a paper by JCB MacKeand and MA Cross, the proposition was put forward that what had been transmitted was a rich mixture of radio frequency energy, a portion of which constituted high-frequency harmonics of the fundamental spark signal. The comparatively low frequency of the fundamental spark signal would have been unlikely to have traversed the Atlantic, but the high-frequency harmonics would have been reflected from the then undiscovered ionospheric layer. By this means the signal was detected at St Johns. MacKeand and Cross contended that the characteristics of the detector used at the receiving end were also of particular relevance.

Poldhu spark transmitter (MAR)

Initially a conventional coherer had been used, without success, and a different form of this device, a 'self-restoring coherer', had then been substituted. The self-restoring coherer employed a globule of mercury against which a steel roller was set and, according to MacKeand and Cross, was a true diode. This was able to respond to the high-frequency pulses of energy despatched from the Poldhu transmitter whereas the rather insensitive filings detector had remained silent.

At the time this experiment received a high level of public acclaim, despite the voices of sceptics, but its importance lay in the future, when the existence of the reflection layer in the ionosphere became known. This opened the way to long-distance wireless communications.

The wireless station at Poldhu was to achieve long-term importance as a research centre at which the potential of short-wave high-frequency radio energy was explored and demonstrated. However, before that was to occur, a series of significant military actions involving wireless was to occur, culminating in the First World War, as described in Chapter 3.

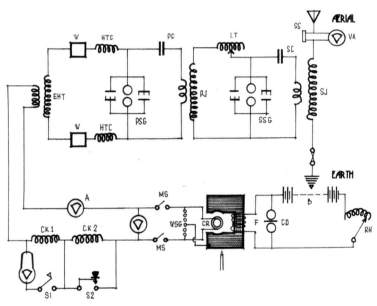

A	ALTERNATOR FRAME	SSG	SECONDARY SPARK GAP WITH PROTECTORS
CR	ALTERNATOR COLLECTOR RINGS	SJ	SECONDARY JIGGER
B	BATTERY	CK 1, CK2	SIGNALLING CHOKES
RH	RHEOSTAT	S1,S2	SIGNALLING SWITCHES
CD	CARBON DISK PROTECTOR ACROSS FIELD WINDINGS	EHT	TRANSFORMER 2000 - 20,000 VOLT RATIO
F	ALTERNATOR FIELD WINDING	HTC	HIGH TENSION INDUCTANCE
WSG	WURTZ SPARK GAP PROTECTOR ACROSS ARMATURE	WW	WURTZ ARRESTORS
MS	MAIN SWITCH	PC	PRIMARY CONDENSER
V	VOLTMETER	PSG	PRIMARY SPARK GAP
A	AMMETER	PJ	PRIMARY JIGGER
LT	ADJUSTABLE TUNING INDUCTANCE	VA	AERIAL AMMETER
SC	SECONDARY CONDENSER	SS	AMMETER SHUNT RESISTANCE

Original transmitter at Poldhu. Larger version on page 318.

2   War Wireless Before 1914

The potential benefits of wireless to mobile combatants had been appreciated almost from the birth of the new technology. The presence of representatives of the British Army and Navy at Marconi's demonstration on the Salisbury Plain would lead to both those services investigating and introducing wireless at an early stage. More ominous in the longer term, was the presence of Professor Slaby from the German Charlottenburg Institute (now the Technical University of Berlin), which led inexorably to the German Army's use of wireless during the First World War.

At the very beginning of the twentieth century, however, a colonial war was to see wireless introduced as an adjunct to conventional military operations, initially as part of the infantry operations in South Africa and then as part of the naval operations and blockade that was mounted against the Boer republics of Transvaal and Oranje Vrijstaat.

The South African War

During the last years of the nineteenth century, tensions between the descendants of Dutch immigrants to the southern part of Africa, the Boers, and Great Britain as a colonising power, progressively increased. Fuelled by the commercial and personal ambitions of a small group of British adventurers in Johannesburg, including the arch-adventurer Cecil Rhodes, outright warfare was the unfortunate outcome of a series of unsuccessful negotiations. Suffice it to say that propelled by commercial greed and political ineptitude, a clash between the imperial might of Great Britain and the determined Boers was to be a dismal start to the new century.

What that first war of the twentieth century was to presage was the increasing impact of new military technology on the conduct of warfare—an impact that would make obsolete the knowledge gained over the previous 100 years. This included the terrible lessons of the American Civil War, which conventional military minds had so singularly failed to grasp.

What the Boers had clearly appreciated was the value of the shovel and of trench building. These two elements allowed the creation of a defensive position that was almost impossible to overcome when coupled with the power and accuracy of the modern breech-loading rifle and, later, the machine-gun. In a historic sense this was particularly ironic because in New Zealand not many years before, during the Maori Wars of the 1860s, the Maori had demonstrated how extremely effective earthworks were in providing high-quality defence against rifle fire and had given the British Army, which had been sent to protect the civilians, a very torrid time.

What the Boer War was also to demonstrate with appalling clarity was the capacity of an intelligent and well-armed guerrilla force to counter the military might of a conventional, large modern army, in the main consisting of infantry. In the terrain over which the Boer War campaigns were fought, the impact of a rifle which had lethal capabilities up to a range of 2 kilometres or more was quite unexpected. Long-range sniping became a devastating element of this guerrilla-style warfare for which initially the British generals had no effective answer. Many war graves in South Africa were the result of the sniping capabilities of the Boer forces, honed by a largely rural existence. In fact, the skills they demonstrated were comparable with those of Australian military recruits in the late nineteenth and early twentieth centuries, and so it is not surprising that the Boer War was to see a significant component of Australian soldiers imported to assist in the British campaign. Only a few years later, this same source would be tapped to provide combatants in the Gallipoli campaign and in Mesopotamia.

Wireless in the Boer War

By comparison with the impact of the modern rifle in South Africa, the first application of wireless communications in an infantry support role produced generally poor results. This was somewhat surprising because, superficially, a method of communications unconstrained by the need to lay conventional telegraph cables should have been of great benefit. However, the British forces included a very efficient unit of line-laying signallers. Following along behind the infantry, the Royal Engineers signallers rolled out telegraph wire from their cable drums and provided a telegraphic service that allowed contact to be maintained between senior officers over long distances. In Thomas Pakenham's history of the Boer War he describes the arrangements of Field Marshal Roberts in the field, noting that:

[Roberts] sat for hour after hour, writing reports and receiving or despatching telegrams by way of the mobile telegraph line unrolled behind a special cart. Its Morse key was one of the keys to the whole campaign; day or night, it gave him the ear of Lord Lansdowne in London, as its Boer counterpart gave General Cronje the ear of Presidents Kruger and Steyn.

In the vicinity of field headquarters, long-distance communication was supplemented to very good effect by the heliograph, which proved particularly useful in an environment in which high-intensity sunlight was consistently available. Set on hilltops, the heliograph was capable of transmitting Morse code messages over distances of 20 kilometres or more.

While wireless communications had seen its genesis in Great Britain and had been applied to naval activities within a very few years, as it turned out it was the Boers who in 1899 first sought to apply the new technology to the anticipated hostilities. Contact was made with the Siemens & Halske organisation in Berlin, which had developed wireless apparatus with a claimed range of 10 to 15 kilometres, and six sets were purchased in August 1899, including the necessary 120 foot (36.6 m) high masts. This equipment was intended to be used for communications between Pretoria, the capital city of the South African Republic founded by the Boers in the Transvaal, and outlying fortifications on surrounding hilltop locations. The

Siemens coherer receiver as supplied to the Boer forces, 1899 (RB)

government had been advised by CK van Trotsenburg, General Manager of Telegraphs, that at £110 Sterling per set the equipment would be considerably less expensive than laying cable to the same locations. This had been anticipated to cost in the region of £9,000 Sterling.

Unfortunately for the Boers, by the time the equipment was shipped, war had already broken out—on 11 October 1899—following the refusal of the British government to withdraw troops from the borders of the two republics as demanded in an ultimatum from President Paul Kruger. On its arrival at the port of Cape Town, the shipment was intercepted and confiscated by customs officials and was then acquired by the British Army.

At the same time, Marconi had been pressing the British government to make use of wireless to support the military and naval forces that had been despatched to South Africa. The War Office response was to enter a six-month contract for the supply of five sets of Marconi transmitters and receivers as well as men to operate the apparatus. It was anticipated that the equipment would be used in association with naval operations and would be installed on naval vessels. However, on its arrival in Cape Town in late November 1899, the War Office decided that a different operational role was appropriate and the wireless stations should be attached to the infantry columns that had been sent inland under General Buller to relieve Ladysmith, then under siege. Captain JNC Kennedy, who along with Major Carr had attended the Salisbury Plain demonstration, was present in South Africa as a member of the Royal Engineers (Telegraphic Section) and was appointed to assist the Marconi engineers to modify the wireless apparatus to suit its new role. This included obtaining Australian sprung wagons to replace the unsprung wagons available in Cape Town which were completely unsuitable for transporting relatively fragile equipment.

Captain Kennedy examined the confiscated Seimens & Halske wireless apparatus. The main difference was the absence of a metal case for the receiver in the Siemens & Halske equipment as a means of overcoming external interference. Kennedy decided that the transmitting elements of this apparatus, as well as the Morse keys, were more substantial than those supplied by Marconi, so they were incorporated into the Marconi apparatus. In principle this should have ensured success but, as identified by Brian Austin in a 1995 article on the subject, three major problems manifested themselves and conspired to make the military attachment a failure.

Creditable Failure

Because it was anticipated that the Marconi wireless installation would be established on naval vessels, no antenna masts had been supplied and

30 foot (9 metre) bamboo poles were used as substitutes. Unfortunately, the heat and low humidity on the Veldt over which military movements were taking place caused the bamboo to split and crack, and all efforts to arrest the problem failed. As a vital element in setting the frequency characteristics of the transmitters and receivers, the poles were replaced by linen kites (similar to the kites used in the first trans-Atlantic experiment in December 1901). Because of the prevailing highly variable wind conditions, this was not a success either, and for 50 per cent of the time wireless contact was not possible.

The Army was less than impressed, and Marconi was rather less than politic in defending his apparatus. His criticism of the failure of the military authorities to make appropriate preparations was evidently very ill received by the Director of Army Telegraphs, whose response was to order the dismantling of the Kimberley mobile stations and the return to the Marconi Company of the apparatus that had been sent to Natal for use in the campaign under the command of General Buller.

The second issue that had a severely detrimental impact on the operation of the apparatus was the frequency of lightning strikes in Natal at that time of the year. Multiple lightning strikes occurred on a daily basis, each producing intense pulses of static electrical noise. This had the effect of triggering the coherers used in the receivers into a permanently short-circuited state which the tapper mechanism was unable to overcome.

The third problem related to the apparently simple issue of obtaining an adequate earth connection to mirror the elevated aerial suspended on the defective bamboo poles. At the time, Marconi tended to deride this as a minor issue, believing it was most likely a reflection of general ignorance of fundamental radio principles. Within only a few years, however, a poor earth connection came to be understood as playing a major part in reducing output power in an aerial system. This in turn led to the construction of enormous earthing mats consisting of copper wire mesh buried in the ground below the aerial elements, which were subsequently employed in trans-Atlantic radio stations.

Somewhat ironically, Marconi's remarks concerning the Army were ultimately beneficial both to naval operations in the Boer War and to the understanding of the technology of radio frequency energy propagation. The wireless apparatus rejected by the Army was immediately acquired by the Navy and installed in five vessels—HMS *Forte*, *Thetis*, *Dwarf*, *Racoon* and *Magicienne*—which were sent as the Delagoa Bay Squadron to blockade Durban and Delagoa Bay. The masts of the five ships were extended to accommodate the wire aerials required and, with access to the excellent earth provided by sea-water, the wireless was as satisfactory as Marconi had promised.

Following this highly successful outcome, the Admiralty concluded that wireless would be an indispensable adjunct to future naval operations and in July 1900 an order was placed for the supply and installation of 26 sets of transmission and reception equipment for Royal Navy vessels. A further six coastal stations were also equipped with Marconi apparatus. This step did not escape the notice of a distant nation also heading towards a military confrontation involving naval forces: Japan. As described in Chapter 2, Japan and Russia came to blows in 1904; and in this confrontation, the wireless apparatus used by the Japanese Imperial Navy was to prove of significant value to the Japanese victory.

Within only a few years, improvements to the wireless apparatus by Marconi and others led the British Army to accept the military value of mobile communication facilities, and portable apparatus was introduced for use by the cavalry brigades. This was followed by progressive integration of mobile wireless facilities up to the outbreak of the First World War in 1914.

Wireless Apparatus: British and Boer

In 1899, wireless apparatus remained little more technically advanced than the experimental apparatus of 1896. Typical of this period just after the start of the Boer War is the Marconi installation illustrated opposite, which consisted of a large induction coil and spark gap, a coherer and tapper, plus a relay and headphones. The other major components included an array of capacitors (condensers or Leyden jars), and the aerial and earth.

An understanding of syntony, or tuning, had led to the introduction of the jigger, an impedance matching device. This can be seen attached to the cabin wall above the Leyden jars and enabled a more efficient transfer of radio frequency energy to occur between transmitter or receiver and the aerial in use. This device was described in a patent (No. 7777) in April of 1900.

Russo-Japanese War

Very soon after the cessation of hostilities in South Africa, in which the British Army had prevailed over the Boer forces, another conflict broke out

Marine wireless apparatus on SS *Minnetonka* (MAR)

in the Far East. The colonial and expansionary ambitions of the Great Bear, Russia, for territory and a 'warm water' port providing year-round access to the Pacific collided head-on with the interests of the emerging eastern power of Japan.

The Russian Empire, stretching across the whole of northern Asia and, via the Trans-Siberian Railway, linking Poland in the west with Vladivostok in the east, reflected a colonising power of huge dimensions. Propelled by the ambition of Tsar Nicholas II, it sought to expand and consolidate its territory in northern Manchuria. Following the end of the Sino-Japanese war in 1895, a warm water port was achieved through negotiations with the Chinese. Territory on the Liaodong Peninsula was leased to Russia, and Port Arthur was developed as a new point of access to the Russian Empire, having major strategic importance in both naval and maritime trading terms. Its development served as a major irritant to Japan, and destroyed the possibility of cordial relations with Russia. Now a rapidly expanding industrial power, Japan saw its growing colonial ambitions directly in conflict with Russian expansionary movements.

In February 1904, the Japanese Naval Fleet under the command of Admiral Togo launched a devastating torpedo-boat attack on the Russian Far Eastern Fleet lying at anchor at Port Arthur, badly damaging the two largest battleships and a cruiser and, some hours later, formally declared war on Russia. There followed a series of land battles and Port Arthur was besieged. Much to the surprise of the Japanese attackers, the Russians raised the white flag and handed over the port and all its facilities on 2 January 1905.

Japanese Naval Victory with Wireless

Prior to this humiliating event, a new Russian Far East Fleet had been assembled in the Baltic Sea under the command of Admiral Rozhestvensky and in October 1904 had set sail for Chinese waters. Following the shelling of British vessels near the Dogger Bank in the North Sea, an attack apparently based on mistaken identity, the Russian fleet was forced to sail around the Cape of Good Hope, rather than through the Suez Canal, to reach the Pacific, a voyage of 33 000 kilometres. During the lengthened voyage Port Arthur fell and Vladivostok had to be substituted as the fleet's ultimate target. To reach this destination meant sailing either to the east or west of the Japanese islands.

The shorter western route via the Tsushima Strait, between Korea and Japan, was chosen and it was here that the wireless facilities of the Japanese Grand Fleet were to produce conclusive and devastating consequences for the Second and Third Pacific Squadrons of the Russian Fleet. Despite specific efforts to pass through this dangerous stretch of ocean at night without navigation or other lights, a Russian vessel was observed by a Japanese naval vessel. A wireless message to headquarters alerted Admiral Togo, and there followed on 27–28 May 1905 the Battle of Tsushima, a battle that has been compared with Trafalgar. The Japanese executed the same tactic as had Lord Nelson in defeating the French. Admiral Togo arranged that his line of battleships should 'cross the T' of the advancing Russian vessels, which ensured that the ensuing Japanese broadside could only be countered by forward-facing Russian guns. Remarkably, Togo was able to repeat this form of attack twice more, totally outmanoeuvring the much less agile Russian vessels.

The continuous broadside from large-calibre Japanese guns on the advancing line of vessels was devastating, and the Russian fleet was almost totally destroyed. Eight battleships and a number of other smaller vessels were sunk, with 4380 sailors killed and 5917 captured, including two admirals, with 1862 interned. The Japanese fleet lost only three torpedo boats and 116 men.

This humiliating defeat is thought to have hastened the revolution that toppled the Tsar in 1917. The destruction of the myth of European invincibility was to support the rise of Japanese militarism in the 1930s that ultimately led to the entry of Japan into the Second World War.

Japanese and Russian Wireless

In 1897 Japan had two naval vessels under construction in London. It had been intended to obtain Marconi's wireless communications system for the new ships but the proposed cost of the equipment and the licence fee proved an insuperable barrier, and the Japanese Navy turned to its own experts. In the event this was probably a very useful decision because wireless was in a state of rapid development and the historical practice in Japan of copying Western technology was in this instance very fruitful.

The task of creating a Japanese version of the Marconi wireless was handed to a task force in which the technical expertise of Professor Kimura of the Second High School in Sendai (now Tohuku University) was coupled with that of Matsunosuke Matsushiro of the Ministry of Posts and Telecommunications (MPT). The result of this collaboration was the creation of the Model 34 wireless communications system that was able to meet the design requirement to span 80 nautical miles (150 km), a distance comparable with the results obtained by the British Navy in 1899. Illustrations of the mechanical and electronic elements of the Model 34 system reveal more than a passing resemblance to the Marconi coherer and tapper mechanism—an early example of reverse engineering. As the naval engagement and subsequent victory of the Japanese Grand Fleet were to demonstrate, the Japanese had created a very impressive analogue of the British Navy wireless system, and one that was certainly more effective than the apparatus installed in the Russian ships.

Although the Japanese success was not something that Marconi and his company were able to share in directly, it appears that the failure of the rival system employed by the Russian Navy produced considerable satisfaction in Chelmsford, the home of the Marconi organisation. The Russian system had been supplied and installed by Telefunken, the Marconi Company's great German rival.

Balkan Prelude to World War

In the short time remaining before the start of the First World War, Turkey was confronted by the forces of Greece and the Balkan states of Bulgaria, Serbia, Rumania and Montenegro in seeking to achieve liberation from its control. The Balkan War of 1912 was notable in the history of wireless communications because, compared with its use in the Boer War not much more than ten years earlier, the new communications technology was to play a significant part, although still less significant than in later wars.

The army of Rumania was equipped with fourteen 1.5 kilowatt Marconi wagon-pack sets and in this was considerably better prepared than the British Army when it crossed the English Channel just two years later. The Ottoman Army was also equipped with wireless, although possessing only one mobile installation.

When the allied armies of Greece and the Balkan states had successfully driven the Turks back to Istanbul, a part of the Ottoman Army was cut off at Adrianople and besieged. By means of the mobile Marconi wagon station, the defenders of the city were able to maintain contact with army commanders in Istanbul for the duration of the siege, which ended with victory for the Bulgarians.

Before the Guns of August: Wireless Developments

In the ten years before the outbreak of World War One in August 1914, radio communications underwent progressive integration into military and naval activities. However, two events that were to have far greater significance for the development of radio technology were merchant marine disasters in which radio played a conspicuous part. A brief description of these disasters is a necessary prelude to consideration of longer term consequences, foremost among them issues that would impinge directly on the application of wireless to warfare and included the radio frequency at which wireless would be used.

The two disasters were the collision of the SS *Republic* and the SS *Florida* off the island of Nantucket, Massachusetts, in 1909 and, far better known, the collision of the RMS *Titanic* with an iceberg in the North Atlantic in 1912. On both occasions the Marconi wireless apparatus was the basis of saving very many lives although, in the case of the *Titanic*, far fewer were rescued than might have been the case if an adequate number of lifeboats were available. Had the master of the SS *Californian* not chosen to roll over in his bunk and ignore a report of distress signals sent up by rocket, the loss of life might have been very slight. In the event, it was the wireless operator of the SS *Carpathia* who heard the distress signal from the *Titanic* and the captain of that vessel who changed course to pick up survivors.

In the collision between *Republic* and *Florida*, the good work of the Marconi operator, Jack Binns, ensured that 1690 passengers and crew were rescued when a number of other vessels responded to the distress signal. But when the Marconi operators on *Titanic* sent out an 'SOS' and 'CQD' signal calling for assistance, only the *Carpathia* responded to collect survivors from lifeboats and floating wreckage. Of the *Titanic*'s 2229 passengers and crew, only 712 were saved.

This tragic event was to see the establishment of legal requirements that ensured that a permanent wireless watch was kept by all vessels above a minimum tonnage. It also led to the opening up of the short-wave part of the radio frequency spectrum, based on new regulations designed to avoid amateur operator-induced interference. It is noteworthy that while the single wireless operator of the *Californian* was off duty at the time of the *Titanic* sinking, it was the *Californian*'s captain who was censured by the subsequent Board of Trade enquiry for failing to respond to the rocket signals.

A significant factor in the saving of lives from the *Titanic* was the wireless apparatus carried on the ship, which from a technological viewpoint was a far more potent communications system than had been available for marine operations only ten years earlier. Rather than an effectively untuned, broad band of spark-generated noise to convey the Morse code signal, the *Titanic*'s Marconi system employed a rotating spark generator that was synchronised with an alternating current generator. Compared with the rough, noisy spark signal available at the turn of the century, rotary spark generated a sound that was more musical in character. This related to the spark repetition rate of about 400 cycles per second (as against the earlier rate of 100 to 150 cycles per second) which was set by the rotational speed of the generator shaft that drove the sparking disk.

The illustration overleaf of a typical marine wireless station from 1912 shows the multiple tuner on the left with the magnetic detector mounted on the cabin wall above. To the right is an induction coil for emergency use, and above it the ship's power supply and switchboard. In the enclosure immediately to the right are the rotary converter and rotary spark enclosure (at the lowest level) and above are the high voltage transformer and aerial inductances and aerial matching boxes. The accompanying schematic

Ship's wireless of 1912, rotary spark (MAR)

reveals the relationships between the various parts.

Apart from the improved acoustic quality of the transmitted sound of the rotary spark, power levels had been increased to a considerable extent, and the transmission input power on the *Titanic* was some 5 kilowatts. Now a daylight range of 400-500 miles (650-800 km) was possible and at night time frequently double that distance was possible.

Larger liners on the trans-Atlantic run were thus able to retain contact with either the west coast of the United Kingdom at Poldhu or the east coast of Newfoundland at Cape Race, either directly or relaying via other vessels. This was the more remarkable in that no form of amplification of the received signal was yet possible. The application of Fleming's 1904 invention, the oscillation valve (also called vacuum tube, electron tube, thermionic valve, tube and valve) lay in the future, as did the fortunate discovery of Lee De Forest in 1907 that resulted in the triode, in which an extra electrode permitted amplification to be achieved. Together these advances allowed the next transformation of wireless communication to take place. As discussed in Chapter 3, a world war became the stimulus for the application of this new technology.

In this period, what was in use as the detector of radio frequency energy generated by spark was a device that had no amplification capacity but despite that, was a highly reliable part of the standard Marconi marine wireless station. This was the magnetic detector (affectionately known as the 'Maggie'), whose only serious shortcoming was the need to keep it

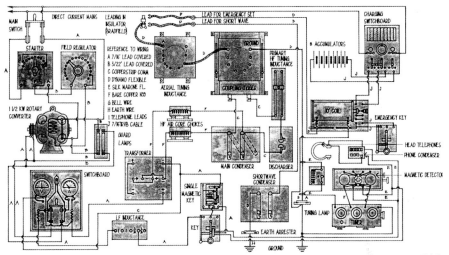

Schematic of typical marine wireless station, 1912 (MAR). Larger version p. 319.

wound up—it relied on a clockwork mechanism for its operation, which involved the rotation of a pair of drive wheels and the movement of a wire rope between the detecting coils to convert pulses of radio frequency energy into audio able to be heard in headphones. The magnetic detector is described in detail, and illustrated, in Chapter 4 (page 87).

Portable Military Wireless

In the military sphere, wireless was developed to provide support to the requirements of the rapidly moving cavalry. With the progressive conversion of this service to reliance on mechanised vehicles, the horse as the prime mover of the army would shortly disappear. Control of armoured vehicles in the battlefield and, ultimately, the tank, would see mobile communications become essential, a service that only wireless could provide.

In the pre-World War One period, what is immediately apparent is the relative simplicity of military equipment compared with equipment developed concurrently for the marine and naval environments. In 1912, the Marconi Company produced a new form of portable wireless equipment designed for land-based military operations that in physical terms was very little different to the apparatus of 1896. The most significant difference was the use of a crystal of carborundum as the detector.

In an article in the August 1912 issue of the *Marconigraph*, forerunner to *Electron-ics World*, a portable wireless telegraphic apparatus, referred

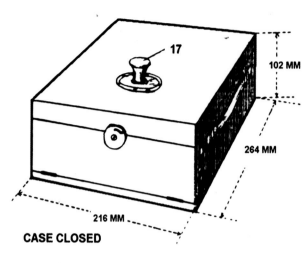

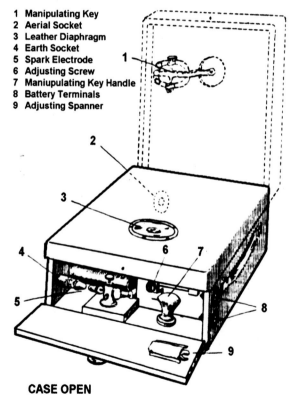

1 Manipulating Key
2 Aerial Socket
3 Leather Diaphragm
4 Earth Socket
5 Spark Electrode
6 Adjusting Screw
7 Maniupulating Key Handle
8 Battery Terminals
9 Adjusting Spanner

Knapsack transmitter, 1913 (*WW*)

to as the 'knapsack' station, was described. In its components can be seen all the fundamental elements of portable military communications that continued to apply for many years subsequently.

Interestingly, the author of the article referred to the history of telecommunications and in particular the conduct of the Boer War at the turn of the century and the sieges that occurred at Ladysmith, Kimberley and Mafeking, observing that there could be no excuse for military ignorance of the successful form of Boer resistance in the South African campaign, and its implications. The battle tacticians of the First World War did not heed the lesson, however, and the persistent use of infantry in frontal assaults on entrenched positions able to deliver concentrated rifle and machine-gun fire with the resultant slaughter

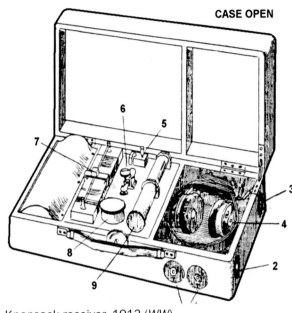

Knapsack receiver, 1913 (*WW*)

remains to baffle the mind of the historian.

The knapsack station consisted of two cases, with the smaller case containing the fixed spark transmitter and the larger case containing a crystal receiver with space for earphones. This latter device employed crystals of carborundum which required a bias voltage to be applied for optimum operation. This was set with a sliding potentiometer (label 6 in the illustration on the left).

The wireless station required four men to carry the component parts—the two wireless boxes (transmitter and receiver) carried by hand, the telescopic antenna mast and battery box, also carried by hand with guying elements and aerial wire drum carried in a knapsack, and, lastly, the rolled-up earthing mats. The apparatus was claimed to have an effective range of 10 miles (16 km) and despite the extensive use of aluminium tubes in the construction of the portable antenna involved loads of approximately 20 to 30 pounds (9 to 13.5 kg) for each of the four bearers. This was a substantial load to carry in addition to normal military equipment and weapons.

The *Marconigraph* article describes the apparatus thus:

> The transmitter, which consists of an ordinary ignition coil requiring a pressure of 6 volts, is contained in a square wooden box weighing 11 pounds. The receiver consists of the ordinary carborundum receiving circuit with jigger, tuning condenser, and four dry cells which may be switched on and off as required. This is also contained in a box which weighs 6 pounds. No elaborate system of syntonisation is provided, this being unnecessary owing to the short wave-length employed, which is so widely different to that in ordinary use as to make the station practically immune from interference.

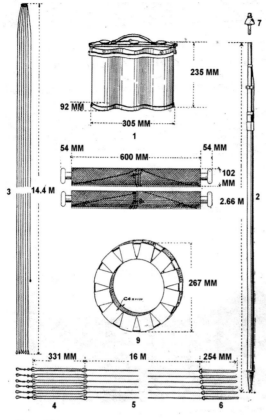

1 Battery 2 Telescopic Mast 3 Aerial 4 Aerial Tension Insulators 5 Aerial Stays
6 Anchor Pegs 7 Mast Insulator 8 Earth Mats 9 Aerial Drum

Aerial mast and components (*WW*)

Only a year later, concerns that future hostilities were likely to require a military response involving wireless were reflected in an article that followed on from the above material. In its first issue of April 1913, the successor to the *Marconigraph*, *Wireless World*, indicates the extent of preparation for major hostilities that had occurred by that time. In an article entitled 'Wireless Telegraphy in the Field', the array of ground-based wireless apparatus available to the army was described:

The most powerful station in general use is the standard automobile station, which is entirely self-contained in one unit, and consists of a six-cylinder 20 h.p. chassis, to which is limbered a two-wheel trailer. The chassis carries the whole of the apparatus in a limousine body. The larger portion of the sending apparatus is fitted underneath the operating bench, but receiving apparatus is fitted above the bench. The dynamo is driven from the engine of the motor-car. The trailer carries the whole of the mast equipment consisting of two 70 ft. masts together with their stays. If necessary, the mast and gear can be carried on the chassis, but the addition of a trailer give the advantage of reducing the weight on the tyres.

The article also describes a lighter form of wireless apparatus designed to be carried on pack animals and referred to as the 'pack set':

A lighter set for use with pack animals has been designed, termed the pack set.

The pack set is arranged to be carried on the backs of four horses or mules, and can be erected in from fifteen to twenty minutes by a squad of eight men. The pack set has a normal range of 50 miles over average country.

From the foregoing it can be seen that despite considerable misgivings among senior military officers relating to the capacity of the enemy to intercept wireless messages, after 1900 the Marconi Company had bent its research and development efforts to preparing for the military use of wireless. In just over a year from the publication of the article in the first issue of *Wireless World*, the world would face the eruption of war following the assassination of the Archduke Ferdinand, heir to the Austro-Hungarian Empire, at Sarajevo in quarrelsome Serbia. His death on 28 July 1914 was the spark that would detonate an explosion of military fury in Europe and precipitate the full-scale battles of the Great War, World War One, which in the end involved the mobilisation of some 60 million citizens.

Anticipating the Tank

During 1911, as the clouds of war were rapidly gathering in Europe, in faraway Australia an enterprising young engineering draughtsman with the rather improbable name of Lancelot Eldin De Mole turned his mind to the problem of moving infantry over rough ground. While undertaking fieldwork in Western Australia he came up with a design for a revolutionary tracked vehicle which to the modern eye bears a remarkable resemblance to what was later developed by the German army and launched as the Panzer tank.

De Mole sent his design drawings to the British War Office in 1912, to receive the patronising response that it was no longer experimenting with tracked vehicles. As events unfolded over the next few years, this was either an example of deliberate evasion or simply bureaucratic stupidity of a very high order. As can be seen from De Mole's drawings and later working model, his invention was a prototype of the modern tank. Moreover, his design appeared a good deal more useful than even the Mark V rhomboidal tank developed by the British Army to respond to trenches and barricades in the middle years of the First World War. The tank, developed in conjunction with wireless communications, was to become one of the key military weapons of the twentieth century and remains highly relevant to battlefield tactics in the present time.

With the benefit of hindsight, and the advantage of access to De Mole's drawings, as well as pictures of his model of the Travelling Caterpillar Fort, as he dubbed the new weapon, it can be seen that a formidable device was ready for development rather than ignominious burial in the archives of the British Civil Service.

De Mole's Travelling Caterpillar Fort, designed 1911 (PRJ)

Records of the development of the British 'Little Willie', 'Big Willie' and 'Mother' prototype tanks suggest that De Mole's ideas may well have had some impact in Whitehall, despite the initial rejection of his proposals and the absence of any later acknowledgement of their existence. Whether or not there was any connection, drawings of the Travelling Caterpillar Fort reveal a device that, particularly in front elevation, bears a striking resemblance to much later tank design. As was common in later tanks, in the De Mole vehicle the tracks are enclosed within the superstructure and the main hull is elevated well above the ground to avoid the problem of 'grounding' experienced by so many of the early British tanks on the Western Front. Thus, through bungling or deliberate suppression, an excellent approach to a war-winning weapon was thwarted and many soldiers would perish who might well have survived if they had been able to cross No-Man's-Land in De Mole's armoured fort.

The model which was constructed to support De Mole's approach to the British War Office is now stored in the Mitchell repository of the Australian War Memorial in Canberra, where one can examine this remarkable piece of machinery at close quarters. Close up, the tracks and bogies show a remarkable resemblance to tank design that came out of the Great War. The Travelling Caterpillar Fort featured a method of steering by displacing the tracks into a curved configuration that might well have produced a far more comfortable change of direction than heaving on lever-driven clutch

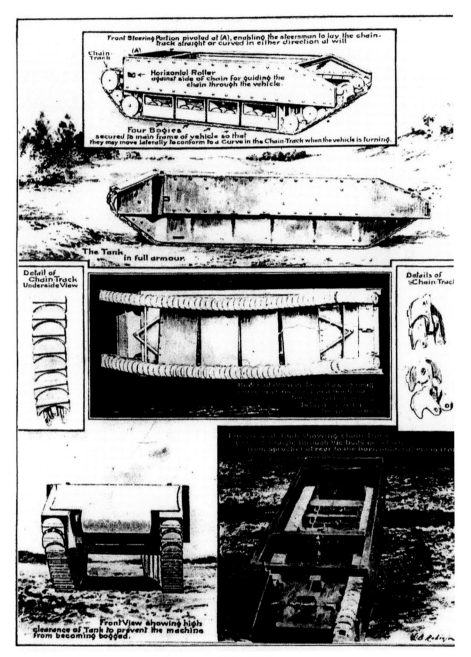

De Mole's Travelling Caterpillar Fort, 1911 (ILN)

controls of the tracks on each side of the tank, as used in first-generation British tanks.

The accompanying illustration shows details of the side of the De Mole model that has been stripped of its covering to expose the tracks and bogey wheels.

The following detailed view of the front end of the De Mole tank shows the clear resemblance to a modern tank, which also features a high clearance intended to avoid 'hanging up' on ground-level obstructions. With the notional addition of a rotating gun turret, the resemblance to much later tanks, up to and including those used in the Second World War and afterwards, is strikingly clear. The image of a relatively modern battle

De Mole tank model with side armour removed (PRJ)

Front end of the De Mole tank (PRJ)

Post-World War Two battle tank (PRJ)

tank from the post-Second World War era helps to make this clear. Only the absence of the rotating gun turret separates the two designs, which are now a full century apart.

Last Steps to War

The Serbian assassin, Gavrilo Princip, had achieved his aim and the heir to the Austro-Hungarian throne lay dead. The tinder resulting from ten or more years of national feuding and a military and naval 'arms race' was set aflame in a war that would last another four years—or perhaps more, if one is inclined to see the Second World War as in reality an extension of the First.

In response to the inaction of the Serbian government over the assassination, the Austro-Hungarian government issued an ultimatum. When Serbia failed to respond to this, the Austro-Hungarian Empire, in conjunction with its allies, Germany and Italy, proclaimed that a state of war existed. This embroiled Russia in the war as an ally of Serbia. Soon Russia's partners in the Triple Entente, France and Great Britain, were also involved. This in turn led to Great Britain's far-distant dominions becoming involved.

Wireless became a vital war weapon that would see battle almost as soon as the 'guns of August' discharged their first furious fusillade in that fateful year of 1914. This use of wireless in warfare on a grand scale heralded an explosion of invention and development that in a few short years would see Marconi's noisy spark completely replaced by the silent radio frequency oscillation of Fleming's valve as modified by De Forest. It would also prepare the ground for the development of radio broadcasting and television that would prove so significant in later twentieth-century conflicts.

3 The First World War

During the 1920s and 1930s, when veterans of the First World War looked back to their experiences on the Dardanelles and the Western Front, the most common perception of the conflict was that it had been the 'war to end all wars'. This vastly over-optimistic outlook was destroyed only a few years later with the start of another world war in 1939. Above all else the First World War was a truly hideous episode in the history of warfare, in which hapless Allied troops were forced to abort an attempted invasion at the Dardanelles and later become hopelessly bogged down in the trenches and mud of northern France and Flanders. Four long years later, when the combatants had beaten themselves into a state of general exhaustion generated by the process of attrition and slaughter, at last an armistice was called, and so the fighting stopped.

The causative factors of the First World War have been probed and analysed endlessly. In brief, the most significant issues related to the perceived need to equalise military and naval power between neighbouring states in Europe, and particularly between Germany and Great Britain. In the politically charged atmosphere pre-1914, various alliances and treaties were entered into that ultimately dragged almost the whole of Europe into the conflict. Led by Germany with the Austro-Hungarian Empire and Italy as partners, the Triple Alliance also drew in the Ottoman Empire of Turkey. This alliance was countered by the creation of the Triple Entente, which included France, Russia and Great Britain and, based on the colonial alliances of the latter, Canada, Australia and South Africa. Later the United States was also to join the conflict. It was the injection of US manpower that finally tipped the military balance in favour of the Triple Entente and saw the end of the war on the 11 November 1918, confirmed by the signing of the Armistice in a railway carriage in the Forest of Compiègne, north-east of Paris.

Twenty and more years later, on 22 June 1940, that same railway carriage in that same forest would be the stage on which the German Chancellor, Adolf Hitler, would confirm the military defeat and humiliation of France early in World War Two. France was forced to accept an armistice that saw it partitioned, with the invidious Vichy administration responsible for the southern part of the country while the northern part remained occupied by the German Army. So it was that a former corporal in the defeated German military forces, who had been 'gassed' in the trenches and bore an ineradicable resentment at the outcome of the war and a long-standing hatred of the Jews of Europe, was to oversee the total humiliation of a supine France.

The End of the Peace

Almost at the start of the twentieth century, on 21 January 1901, the death of Queen Victoria, the longest reigning monarch of the British Empire, signalled the end of an era of comparative peace in Europe. During the following years, that peaceful state was steadily eroded. A subtle but potent factor in the progressive decline in state relations was the jealousy and dislike held by the heir to the German throne, Crown Prince Wilhelm II, towards his uncle, the new British sovereign, Edward VII. That enmity would provide part of the impetus for military and expansionary activities in Germany, in particular attempting to match the naval power of Great Britain.

Wilhelm's plans included the widening of a major strategic waterway, the canal linking Kiel on the Baltic Sea with the North Sea near Cuxhaven on the River Elbe, allowing it to accommodate newly constructed battleships and so move them quickly from the Baltic to the open North Sea. Up until this time, an extended voyage around Denmark was necessary for ships moving from the Baltic to the open sea and the Atlantic Ocean beyond. Construction in Britain of the new Dreadnought class of battleships had been the impetus for German naval competition after 1900 and the undertaking of a strenuous battleship construction program. Expansion of the Kiel Canal became a national goal, enhancement of this important linkage being seen as removing a barrier to war against France and Great Britain. The summer of 1914 was contemplated for the commencement of hostilities—as subsequently occurred.

The assassination of Archduke Franz Ferdinand in Sarajevo set off four hideous years of conflagration across much of northern Europe. Alongside the new weapons of war, the magazine rifle and the machine-gun, marched wireless, which developed rapidly to meet the demands of the combatants. With the creation in 1916 of another new weapon of war, the tank, wireless emerged as the vital ingredient that allowed communications with this new mobile device to be maintained.

First Steps in Communications Warfare

On the declaration of war in August 1914, British naval forces were ordered by wireless to mobilise and assemble in the North Sea. A dramatic early offensive action by the Royal Navy was to have a long-term impact on international communications, during the war years and afterwards. This was the dredging up and cutting of the undersea telegraph cables that connected Germany to the United States and South America, which passed around the southern part of Great Britain and Ireland before crossing the Atlantic. The cutting of the cables had the immediate effect of constraining Germany's long-distance telecommunications to the use of wireless. The principal international station in Germany, located at Nauen, was able to contact South America and the United States, and stations in southern and western Africa. It was not long before even these linkages were broken as well. Kamina in Togoland was destroyed by its German operatives on 24 August 1914 when capture by British forces appeared inevitable. The German Windhoek wireless station was captured in May of 1915 by forces of the South African Union, now a British ally.

While it may seem curious that the major cable networks of European powers with a long-term history of conflict should pass in close proximity to each other's territory, this was very much a reflection of the history of undersea cable installation and the large part that Great Britain and the United States had played in establishing the chain of linkages that spanned the globe by 1914.

That early belligerent act of cable-cutting was to see a German response in locations very far from Europe, initially at Fanning Island in the central Pacific Ocean, where the Vancouver to Fiji cable was attacked by the German Navy in September 1914. The light cruiser SMS *Nürnberg* landed a party of marines who destroyed the cable station but were unable to find the foreshore end of the cable itself. Only a month later, repairs had been completed and the cable link was again operational. The *Nürnberg* was later sunk by HMS *Kent* in the vicinity of the Falkland Islands.

Of particular significance to long-distance communications between Australia and Great Britain were the activities of the German raider SMS

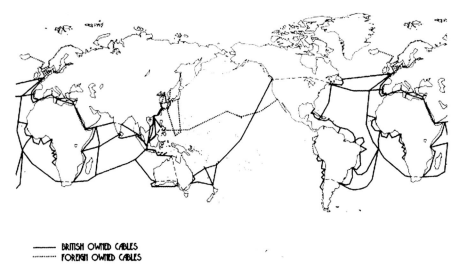

— BRITISH OWNED CABLES
···· FOREIGN OWNED CABLES

The mature undersea cable network (PRJ)

Emden, which attacked a number of cable relay stations in the Indian Ocean. At that time, the Cocos-Keeling Islands were the site of a long distance cable relay station and by 1914 also the site of a wireless station.

Naval Battle at Cocos-Keeling

One of the new German battleships that exited into the North Sea via the Kiel Canal in April 1910 was the recently commissioned light cruiser, SMS *Emden*. Entering service in 1909, the *Emden* weighed 3364 tons and was capable of a speed of 23 knots, or somewhat over 40 km/h. A coal-burning steamship, *Emden* was the last German battleship to be driven by triple-expansion reciprocating steam engines. It carried as its main armament ten 4-inch guns, along with a number of rapid-fire smaller calibre guns and a pair of torpedo tubes.

Under the command of Captain Karl von Müller, *Emden* sailed to the German colonial port of Tsingtao in China where it was based until just before the outbreak of hostilities in August 1914. Müller put to sea just prior to that date, the example of the Russo-Japanese war of 1904 leading him to fear that his ship might be captured. The *Emden* ranged out into the Pacific and Indian Oceans and preyed upon merchant vessels, sinking or capturing a large number in the ensuing months.

The British–Australian cable links were now a prime target and *Emden* sailed to Direction Island, in the Cocos-Keeling group, on which was located the cable and wireless station of the Eastern Telegraph Company. A party of 49 men and two officers landed there under the command of Lieutenant Helmuth von Mücke, with instructions to destroy the cable station and cables as well as the wireless station and its 54 metre mast.

This work was duly undertaken but unbeknown to Captain Müller, the timing of this action was uniquely unfortunate for the *Emden*. Only 80 kilometres to the north-east, the ANZAC expeditionary force was sailing in convoy from Perth to Bombay, escorted by a number of Royal Australian Navy vessels, including the new Town class light cruiser, HMAS *Sydney*. When the *Emden* was first sighted by the Deception Island cable station and wireless station operatives, SOS messages had been sent out, by cable to Electra House in London and by wireless to the naval convoy close by. *Sydney* was promptly deployed to intercept the *Emden* and arrived at Direction Island while the destruction of the cable station was still going on.

At 5400 tons and capable of slightly more speed than the *Emden*, *Sydney* was a somewhat larger vessel and carried eight breech-loading 6-inch guns as well as smaller calibre rapid-firing guns and machine-guns. *Sydney* was a major adversary, and the outcome of the engagement was probably inevitable despite the best efforts of Captain Müller and his sailors.

Müller immediately brought the *Emden* under way, leaving the shore party stranded, while the Australian vessel approached to determine the raider's identity. This was soon confirmed by an opening salvo from the *Emden*'s 4-inch armaments. The response was immediate and quite devastating. Despite the destruction of its ranging apparatus, *Sydney*'s superior armament soon reduced *Emden* to a smoking wreck.

To avoid the sinking of his vessel, Müller ran the *Emden* ashore on North Keeling Island where, finally, after further shellfire from *Sydney*, the white flag was run up and the surrender was given. Captain Müller and his remaining crew were taken aboard *Sydney* to become prisoners of war. On shore, Lieutenant Mücke had commandeered the *Ayesha*, an ancient sailing vessel belonging to the owners of the Cocos-Keeling Islands, the Clunies-Ross family. Without a chart to guide them, Mücke and his shore party sailed to Batavia in the Dutch East Indies, from where they returned to Germany seven month later, to a heroes' welcome.

This highly successful use of early naval wireless communications was followed by the demolition of wireless stations in a number of locations in the German colonial holdings in the Pacific by men of the volunteer Australian Naval and Military Expeditionary Force (AN&MEF). Among these were the station at Apia in German Samoa in late August 1914, and then the station on Nauru in the Marshall Islands of German New Guinea. These were followed by the destruction of Herberthihe on Neupommern Island (now New Britain), also in German New Guinea. After that Australian wireless activities were to involve a campaign on a peninsula far away in the northern hemisphere—the Dardanelles—that helped to create a self-image of Australia as a nation that would persist for the following hundred years.

Wireless on the Western Front

While the naval engagements in the Atlantic and the Pacific were proceeding, in the weeks after August 1914 Germany swept into Belgium and the north of France. Pressing the rapidly mobilised French and British forces back in a helter-skelter drive to the south, the German Army was finally brought to a halt in the First Battle of the Marne, and driven back from the intended target of Paris.

Pressed back onto a line that ran almost from the Swiss border in the south-east to the edge of the English Channel in the north-west, the adversaries froze into a stalemated embrace of mutual annihilation in which the obsolete tactics of a bygone military age ensured that thousands upon thousands would die, and saw very little military advantage achieved by either side over an extended period. Thus was born the Western Front, upon which shellfire, machine-guns and gas descended to ensure that the First World War would be remembered as truly a 'Hell on Earth'.

Although wireless had been available at the outbreak of war, the British Expeditionary Force arrived in France in a woefully underprovided state. One motorised lorry wireless installation from the Marconi Company was there to support the mobile cavalry, together with nine horse-drawn wagon sets. In the new form of trench warfare, the cavalry was made irrelevant and would remain irrelevant until its four-legged mounts were converted into the metal tracks of the first armoured tanks in 1917.

In the meanwhile, with the need to communicate from the front line to headquarters in the rear, wireless provided the vital link. Wireless was invaluable where conventional telephone cables had been torn up by shellfire and where enemy listening devices that could capture unguarded remarks from officers using field telephones had been detected. The British Army's initial reluctance to utilise wireless was soon overcome by the advantages of a communications system that did not rely on vulnerable wires.

By the time that the Battle of the Marne brought a halt to the advance of the German Army, a further ten mobile wireless installations had been provided to the British Army. However, when a new enterprise was proposed to relieve the situation along the Western Front through a campaign in Turkey, for reasons that are now lost in the mists of history additional wireless apparatus was not available to the British Army. Fortunately Australian forces went to war with six Marconi portable wireless stations that were deployed to support that new enterprise: the Gallipoli campaign.

Horse-drawn Marconi wireless wagon set, 1914 (MAR)

Wireless at Gallipoli

In a desperate attempt to break the deadlock of the Western Front, the First Lord of the British Admiralty, Winston Churchill, conceived a surprise attack on the 'soft and vulnerable underbelly' of the German military machine and its ally, the Ottoman Empire, on the shores of the Dardanelles. However, what could have been a highly effective strategy depended first and foremost upon secrecy—and this was the most conspicuous failing of the Gallipoli campaign. In besieging the nearby city of Troy, on the far side of the Dardanelles approaches, the ancient Greeks had been significantly more adept at employing secrecy and subterfuge. Regrettably, at Gallipoli, the subtlety of the Trojan horse tactic was conspicuous by its absence.

By the time the British, French, Australian and New Zealand forces were about to land on the beaches around the headland of the Gallipoli Peninsula, the defending Turkish forces were about as well advised of the impending attack as it is possible to imagine. Not only that, the topography of the peninsula seems not to have been adequately investigated, and its rugged and impenetrable nature came as an unpleasant surprise. The Allied invaders were unable to penetrate inland to any appreciable extent and instead were brought to an impasse just as impossible as that of the Western Front.

Although remarkable for its minimal presence in the official history of the Gallipoli campaign, as created by noted Australian historian Charles Bean, wireless went to war provided by the First Signals Troop of the Australian Corps of Signals, First Australian Imperial Force (AIF). This was the result of conspicuous foresight on the part of the Australian government that had resulted in the purchase of six Marconi horse-borne pack sets ('pack' being a military term to describe a device or package that could be transported either by hand, by cart or by horse). Their delivery in 1913 had allowed the development of signals capability in the Australian Army well in advance of British Army capabilities. Two sets each went to New South Wales and Victoria, the fifth to Queensland and the sixth to South Australia.

The two pack sets for Victoria went to the Australian Army Corps of Engineers and were put into the care of Sergeant OF (Orm) Metcher, with the operation of the wireless in the hands of Sapper HD (Bert) Billings, and Sapper W (Bill) Dobbyns in charge of the motor generator set. It is fortunate

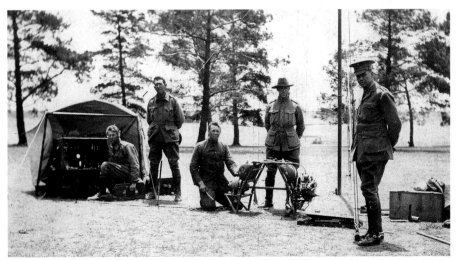

Marconi pack set at Broadmeadows, Victoria, 1913 (RAS)

that Bert Billings was one of the persons instrumental in introducing this new Marconi wireless apparatus into the Australian Army—he appears to have had a phenomenal memory and, very late in life, his recollections of this period were transferred to paper and provided to the Australian War Memorial in Canberra.

As described by Billings, who already held the experimental licence XJP to operate wireless apparatus, the pack set was an assemblage of parts that added up to a load of some 4 hundredweight (200 kg). In addition to the four packhorses needed to carry the various parts and the six signallers required to set up and operate the wireless station, a Signal Troop was supported by a general service limber, drawn by a team of four horses and accompanied by two drivers and a brakeman.

Billings' description of the elements of the wireless station is of interest because a fully trained and experienced signal troop was able to erect the masts and aerials and be operational in under five minutes. The station consisted of the following items:

- The receiver cabinet containing carborundum detectors and bias battery
- A box containing cylindrical high voltage glass Leyden jar capacitors
- An alternating current transformer—100 volts AC to 25,000 volts AC
- 500 watt spark transmitter
- 2¾ horsepower Douglas twin petrol engine and AC alternator generator
- Two 30 feet (10 metre) high tubular steel masts consisting of 5 foot long sections
- 100 metres of aerial wire
- Ropes and guys for the two masts
- Earthing mats made of copper mesh
- Petrol and oil as well as spare parts

The pack set wireless station had a guaranteed range of 30 miles (50 km) or more, although even during daylight hours that was frequently exceeded by up to three times. At night the range was even greater, and between 200 and 300 miles (320–500 km) could be achieved. The equipment was operated at wavelengths between 300 and 1000 metres, although 700 metres was the usual operational value. This represents a frequency range of 1 megahertz to 300 kilohertz and lies at the bottom end of the AM broadcast band (where broadcast stations of the ABC are located in the Sydney region). Messages were sent in five-letter code and plain language communications were prohibited.

Four wireless packs awaiting horses (MAR)

Marconi pack set (MAR)

Soon after the declaration of war, Australia offered a contingent of 25 000 men. This party was despatched to Egypt to provide support in the campaign against the Ottoman Empire. The recently created Signal Service of the Royal Australian Engineers went with this contingent and on arrival in Cairo were attached to the First Light Horse Brigade. From here, the First Signal Troop was sent to support the British Army in its attack on Gallipoli. Rather than landing at Anzac Cove with the AIF contingent, it was landed on Cape Helles with 500 men of the Essex Regiment of the 29th Division of the British Army in the assault on 'W' Beach.

Here Turkish counter-attacks prevented a permanent wireless installation and two days later, the signallers moved to 'V' Beach where contact with the Royal Navy was established and messages were passed relating to spotting the fall of shells from the naval vessels. The wireless station was set up in close proximity to the old Sedd-al-Bahr Fortress and received messages from a forward observation post via telephone landline for transmission to naval vessels moored offshore.

Thus, in the absence of wireless apparatus available to the British Army, in this initial period of the Gallipoli campaign it was the twelve men of the First Signal Troop, AIF, who provided a 24-hour wireless link to the Royal Navy. The contingent was subsequently relocated to support the French attack on the peninsula from 'S' Beach and, later, to 'X' Beach to support the 6th Gurkha Rifles. Here the Australian signallers were accepted as 'brothers in arms' and formed a very good impression of this element of the British Army and their deadly sharp weapon, the *khukri*.

Some weeks later, when the stalemate on Gallipoli was fully established, the First Signal Troop was moved to the headquarters of the AIF at Anzac Cove. Here communications support was provided to the Signals Office of the First Light Horse and to its commanding officer, Colonel (Later Sir) Harry Chauvel. This contingent had been forced off their horses and precipitated into the trenches with the Australian infantry due to the difficult country over which the campaign was waged, terrain particularly unsuitable for mounted warfare.

Following an abortive landing and attack north of Anzac Cove at Suvla Bay in August 1915, the decision was taken to remove the Allied troops from the peninsula. Then came the problem of achieving this without the Turkish defenders becoming aware of the movement and pursuing the British, French, Australian and New Zealand forces into the sea. Eventually secrecy prevailed, and the manoeuvre was performed in an exemplary manner that avoided any further deaths.

If the abortive Gallipoli campaign had been a terrible waste of British, French and Commonwealth lives, what was now to transpire was unimaginably worse. From Gallipoli, the Australian and New Zealand infantry were transferred to the Western Front to participate in the carnage that would long colour antipodean attitudes to British authority. The Light Horse and the associated wireless troops were sent to a far more open and appropriate location for their type of warfare:

the Sinai and Palestine as well as Mesopotamia and later into Persia (modern Iran).

Signaller at Gallipoli

As a youth of fifteen, Bert Billings became a clerk and telegraphist on the Victorian Railways and by 1912 had obtained a wireless experimenter's licence. When war broke out, Billings applied to join the Army as a signaller and became a part of the 21st Signal Engineers where he learned to use the heliograph for sending messages using Morse code as well as flags.

In a remarkably farsighted move, in 1913 the Australian government had purchased six Marconi 500 watt pack sets. With his previous experience in amateur wireless, Billings was selected as one of the team of signallers given access to the two pack sets sent to Victoria. Initial exposure and operation of this apparatus by Bert Billings, Orm Metcher and Bill Dobbyns was captured by a camera at Broadmeadows, Victoria, in 1913. Coincidentally, this was the same equipment that Billings later accompanied to the Gallipoli Peninsula and operated at Cape Helles.

Billings' Australian War Memorial papers recount what transpired on the first traumatic and chaotic day:

> Everyone on the ship was called at 4 am, but many of (sic) had not slept much, and by 5 am we could hear gun fire from the Royal Navy in the darkness ... as

Gallipoli wireless station—Dobbyns, Metcher and Billings, 1913 (MOV)

the 'Dongala' sailed along in the early hours of this Sunday morning we saw the flashes and the shells burst on the coast, then as it got lighter we saw the warships and the actual bursting of the shells.

Everyone was so engrossed in watching the battle that it was not realised that our ship had drifted right between HMS Queen Elizabeth and the shore, at which they were shooting 15 inch shells as quickly as they could. The first warnings we had were from the sirens of the QE which had to cease firing to avoid hitting us (with 15 inchers!!)

Needless to say, our ship rapidly reversed out of the way, the Captain no doubt having received a severe fright and we certainly did too.

At about 11 am, we were called away from our grandstand view to enter our boats and with all our equipment we were lowered into the water. Two other boats were also lowered with a load of Essex men, and at 11.30 am, a Naval steam pinnace took us in tow, bound for 'V' beach (and our first experience of war).

Half way towards the shore which we could see was still being heavily shelled by the Navy, another steam pinnace raced up to us and we were ordered to go to 'W' beach instead, as 'V' (our intended destination) was still under siege and not in Allied hands.

This change of plan was necessary (and very urgent) as 'V' beach, which proved the most costly of the Cape Helles landings, was still under heavy fire by the defenders who had the 'Munsters' and the 'Dublin' (Fusiliers) pinned down and/or unable to move out of the steamer River Clyde for the rest of the daylight hours.

Both the Essex Regt. and ourselves were very lucky when we were diverted in time from 'V' beach to the 'W' beach that morning.

After our fortunate diversion from 'V' to 'W' our boats were taken in as close as possible and we were then dropped to row ourselves in as close as possible. We managed to get in close near the southern end of the beach near a lot of rocks and a high cliff.

We carried all our equipment (the station alone weighed 4 cwt.) over the sand and sheltered in the side of the cliff under a tarp, which we were fortunate to find, and sheltered there awaiting orders.

That landing had only been taken a short time previously by the Lancashire Fusiliers and a long line of bodies was on the beach, still more still (sic) in the water and huge piles of gear lay everywhere.

Metcher counted 86 bodies (including 3 officers) on the beach, plus many others in the water.

This was my first sight of war, and something I will never forget.

We remained on the beach all day, more or less 'lost', as we could not get to our appointed position on 'V' and there was no room (even if we had be (sic) required) to erect our station where we were, so we just waited for orders to come along.

Things quietened during the day and in the afternoon Orm Metcher and I were searching through an old Turk trench just nearby, in which five Turkish bodies lay. We, and others, were looking for anything of interest when one of the 'dead' men objected to being searched and decided it was time he admitted he was not dead, but only laying 'doggo'. It turned out that he was not wounded at all, but had probably been knocked unconscious by an explosion and came to after his mates had be driven out. He was taken prisoner and was probably quite happy for the rest of the war.

Night came and we huddled under our tarp. still waiting for orders.

Tried to sleep a bit but the Turks made a very heavy counter-attack both against our front and that at 'V' beach. (They had rushed 5000 more men) and things looked bad. At 11.40 pm every man on the beach was ordered to fix bayonets and to go to the top of the cliff into a reserve trench (One of the Turks old ones) and to be ready for a last stand.

As we were not under anyone's definite control we just followed orders and went up top—we both remember a very agitated officer running around and waving a revolver, and yelling for us to get up top ... and that was the end of my first day at Gallipoli.

In the months that followed, the Australian signallers were moved around to support the British Army and later joined the Anzac Cove forces further north. Months later when it was realised that the campaign had reached a stalemate, it was clear that a strategic withdrawal was the only rational option. This was accomplished with great skill and the absence of further battle casualties.

Mobile Wireless in Palestine

By the beginning of 1916, the Western Front was fully established as a largely static moonscape of churned-up mud and shell-blasted tree trunks, dotted with craters and the remains of horses and men scattered where they had been caught by the whirlwind of rifle or machine-gun fire.

The major problem of interception of cable-borne audio voice signals by the German Army had been overcome by the introduction of the Fuller phone, devised in 1915 by Captain AC Fuller of the British Signal Service. However, the destruction of the cables remained an insuperable problem. As the war continued, wireless was introduced and progressively demonstrated its benefits in flexibility and its capacity for rapid redeployment although,

Pack set in the Middle Eastern campaign (BUR)

from the beginning, interception had been a problem that could only be overcome by the encoding the Morse signals. By the end of the war, wireless had become an accepted, if not indispensable, part of the military armoury as static trench warfare transformed into mobile warfare and the introduction of the tank demanded a means of communications that only wireless could provide.

Following withdrawal from the Dardanelles, the Australian and New Zealand forces were split, with the infantry sent to reinforce the troops on the Western Front and the mounted components sent to join the campaign in Egypt and Palestine and later into Mesopotamia and Persia. With the latter went the Australian Army signallers, their horses and their wireless sets.

The campaigns in the Middle East were to benefit greatly from the new mobile communications technology. When cavalry and light horse units moved ahead to attack the elusive Turkish Army, the wireless carts followed along behind, maintaining command and control over distances up to 100 miles (160 km). This was made possible by a process of 'leapfrogging', with an established wireless station being disassembled while another was trotted forward to catch up with the tail-end of the mounted forces. Once the gap was closed, the forward station was quickly set up and the first station in turn trotted forward to overtake the new location and remain in contact with the horsemen.

The primary link from the Mediterranean to India and the Far East, the Suez Canal was inevitably a target by the Ottoman Army, which advanced to take it shortly before the commencement of the Dardanelles campaign in February 1915. The Turks were beaten back to the eastern side and by the time of the withdrawal from Gallipoli, late in 1915, the decision was made to establish Allied forces on the eastern side. This was the start of the Palestine campaign, in which Allied forces pushed to the east and north supported by a new water-supply pipeline to take water from the Nile River to the advancing forces. A rail line was also laid to follow the troops and supply their demands for food and ammunition in the desert region of the Sinai, where all such material had to be provided from stores in Egypt.

Commencing with the successful battle to take the Egyptian town of Romani, somewhat to the east of the Suez Canal, a campaign was now launched to dislodge the Ottoman Army from the historical biblical lands of Palestine. In a series of forward movements and outflanking of the Turkish forces, firstly General Sir Archibald Murray and later his replacement, General Sir Edmund Allenby, drove their opponents from the banks of the Suez Canal through the Sinai to Romani and progressively northwards to Jerusalem, which was taken on 8 December 1917.

In concert with the Arab army under TE Lawrence (Lawrence of Arabia), the Allied forces pushed further northwards, and on 1 October 1918 the Australian Light Horse swept unopposed into the historic town of Damascus. Major Olden advanced into the town early in the morning, where he accepted the surrender of the Turkish forces. Later that day, Lawrence and his Arab forces entered Damascus and Lawrence, without justification, claimed priority in the victory.

At the end of October 1918, the Turkish Army was forced to sign an armistice and two days later it surrendered. With this event, the war in the Middle East came to an end, as did control of the Middle East by the Ottoman Empire, which had held sway for over half a millennium.

Mobile Wireless: Mesopotamia and Persia
In parallel with the campaign in Egypt and Palestine, British and Empire forces had been deployed to eject the Ottoman Army from Mesopotamia (now Iraq). Later the campaign would be extended to Persia (Iran) and on to its eastern edge on the Caspian Sea. This campaign had considerable strategic importance because the area included major oil deposits which remain of critical importance to the present day.

At that time, the oil assets were under the control of the Anglo-Persian Oil Company (which became British Petroleum, BP, in 1954). At an earlier time, the organisation from which Anglo-Persian had sprung had secured the exclusive rights to drill for oil throughout the Persian Empire. This was the source of fuel for the Royal Navy, which from 1911 had progressively abandoned coal-fired boilers for oil fuel and turbine-powered engines in its battleships. In addition to the oil fields in Persia, on the western edge of the Caspian Sea were located further oil deposits and a major oil refinery at Baku, now the capital of Azerbaijan.

An initial campaign commenced in April 1915 under the command of General Sir John Nixon, who was able to take Basra and later take complete control of the territory traversed by the Tigris and Euphrates rivers in southern Mesopotamia. However, later expeditions along both the main rivers led to a disastrous defeat. At Kut-al-Amara in December 1915, Major General Townshend and his forces were besieged and in April 1916 was forced to surrender to the Ottoman Army with 200 British and 6000 Indian troops taken as prisoners of war.

Following the replacement of Nixon by General Sir Stanley Maude, the offensive along the two strategic rivers was recommenced, involving two corps and a total force of 166 000 men. This led to the taking of Baghdad in March 1917 and a further series of engagements along the Tigris, Euphrates and Diyala rivers intended to destroy residual Turkish infantry formations.

During this same period, a highly secret expedition was organised under Major-General Lionel 'Stalky' Dunsterville, intended to prevent the oil fields and refinery at Baku falling into the hands of the Turkish Army as the Russians withdrew from hostilities and departed northwards, following the Russian Revolution of November 1917. An equally important strategic ambition was to prevent the German Army reaching India via the Caucasus in concert with the Ottoman Army.

While conceded by historians to have been something of a 'sideshow' compared to the Western Front, from the perspective of wireless communications, the Dardanelles, Sinai and Palestine campaigns were of major importance, . In particular, the lesser campaign in Persia demonstrated how valuable the new communications technology could be when appropriately employed. Leapfrogging behind the cavalry, or moving along the banks of the broad Tigris and Euphrates rivers, wireless allowed widely separated and highly mobile mounted forces to maintain contact in a fashion that military operations had never enjoyed previously.

In Dunsterville's expedition (codenamed Dunsterforce), wireless helped to support a remarkably small force of not more than 1000 officers and men who were sent to penetrate an ostensibly neutral but unenthusiastic

Persia as far as the shore of the Caspian Sea at Enzali. From this port, Dunsterville was able to sail to Baku and engage in negotiations with residual elements of the Russian Army and its Armenian allies and later with the new Bolshevik Army as a basis for resisting Turkish attack.

Marconi pack set in Persia (BUR)

Wireless station in Persia (BUR)

Unfortunately this effort proved little short of a military fiasco when the remnants of the Russian Army withdrew to the far side of the Caspian Sea and the remaining Armenian forces proved totally ineffective in defending Baku. The Armenians then opened negotiations to surrender to the Turks without informing the British, which led to inevitable withdrawal by ship to Enzeli, successfully accomplished.

Model T Fords with pack set (BUR)

The efforts of the ANZAC Signal Troop were described in great detail in the book *With Horse and Morse in Mesopotamia* (1927), a substantial volume compiled under the editorial hand of Eric Keast Burke. In a lengthy foreword Dunsterville paid tribute to the support given by the ANZAC signallers during the advance into the east of Persia. They had been able to provide a consistently high level of message-handling under trying conditions and Dunsterville was able to say of them:

> Had it not been for the work of the various Anzac stations during the early days of our penetration in to Persia, we should have been entirely cut off from communications with the outer world, and the success of the expedition would have been still further jeopardised ... Throughout the operations their work was carried out so efficiently and unobtrusively that the Command and Staff rather took them for granted ... their sheer efficiency kept them from the prominence they merited.

In the end, the efforts of Dunsterforce were to prove fruitless, with the withdrawal of the troops to Baghdad. In retrospect, and given the turbulent situation in Russia following the revolution and the failure of British High Command to adequately support Dunsterville with a substantial force, this outcome was not at all surprising. After the withdrawal, Dunsterville decided to confront his superiors with the reasons for the failure of the campaign and for his pains he was dismissed from command of the 'secret' force. Dunsterforce ceased to exist in 18 September 1918 and Major-General Dunsterville retired into relative obscurity, surrounded by lingering doubts that 'Stalky' might have managed more with his meagre forces than could have been humanly possible. As many of his comrades were aware, Dunsterville was the youthful companion of Rudyard Kipling and hero of that author's book for boys, *Stalky and Co.*, published in 1899.

Major-General Lionel 'Stalky' Dunsterville (BUR)

Wireless Developments in World War One

What becomes very apparent in looking at the wireless equipment in use during the First World War is the range of the devices created in just four years, and the rapidity with which technology produced increasing power and refinement. This equipment falls naturally into two groups, involving either spark or continuous waves. The array of wireless devices used by the army and air force is set out in a three-part article by Basil Schonland (later Sir Basil) dealing with wireless equipment and operations during the First World War that appeared in *Wireless World* in July 1919.

The British Expeditionary Force arrived in France in August 1914 with 1½ kilowatt Marconi wireless equipment carried on horse-drawn carts and limbers and a motor lorry. On the Western Front this equipment became redundant and recourse was made to landline telephone communications. On the Gallipoli Peninsula, smaller and somewhat more portable wireless equipment was employed with apparent success and later transferred to Egypt where it was used in the Middle East campaigns.

On the Western Front, as vulnerable telephone lines were destroyed

Motor lorry-based wireless (BUR)

by shellfire, in 1916 new and simple wireless equipment was introduced in the form of the BF trench set. During the First Battle of the Somme in July 1916, wireless support provided by the BF trench set demonstrated that where the telephone cables were destroyed, wireless was able to maintain communications and pass messages. This successful service had the effect of starting a general reduction in the entrenched scepticism that had existed in the British Army, and particularly among senior staff officers, relating to this new means of communications, no doubt assisted by the realisation that the German Army had introduced its own system of wireless communications. The German equipment appeared to copy the characteristics of the Marconi wireless apparatus, although it was manufactured by Telefunken.

The BF Trench Set

The BF trench set was about as simple a piece of apparatus as could have been conceived at that period for use by inexperienced front line infantry troops. It has been suggested that the notion that 'any bloody fool' could operate this apparatus was the basis of its designation. However, in official army manuals a more elevated meaning for the two cryptic letters has been ascribed: the British Field trench set is suggested.

The schematic for the BF trench set appears in Chapter 4, on page 85. The transmitter employed an induction coil in conjunction with a simple magnetic interrupter in the Morse key circuit. The set was supplied by a 10 volt, 12½ amp-hour battery and was intended to operate on any one of three wavelengths, 350, 450 or 550 metres. The companion receiver used a carborundum detector with its bias battery and was designed to allow reception over a range of wavelengths between 300 and 600 metres. The complete receiving and transmitting apparatus was contained in a 406 x 254 x 228 mm (16 x 10 x 9 inch) wooden box with a hinged lid and demanded a strong arm to heft its 14 kilogram (31 pound) weight. As with the portable wireless apparatus of 1913, a considerable quantity of ancillary gear was required and a team of three or more soldiers was needed to carry the apparatus into the battlefield and erect the antenna system. In practice, this supporting group might need to be as many as nine in number, a problem that produced a demand for more compact and lighter equipment.

With a wire antenna from 60 to 80 metres in length, and supported on 5 metre sectional steel masts, and a copper earth mat 4.25 metres long that could be rolled up for transport, the BF trench set was capable of communicating over a distance of about one kilometre and in favourable conditions up to four times that. The range achievable was very much a function of the physical characteristics of the terrain and surrounding vegetation.

BF trench set (PRJ)

To set up the carborundum detector in the receiver for optimum audio sound level, a small buzzer allowed the potentiometer to set the correct voltage across the crystal. High-impedance headphones were used in conjunction with the rectifying elements of the receiver. The other part of the equipment enclosed in the wooden case was the transmitter, which was inductively coupled to the antenna. Coupling was varied by means of stepped tapings on the antenna coupling coil and by variable capacitance across the transmitter output coil. The maximum output that could be achieved in this operation was indicated in the antenna current meter.

Wilson Sender (Transmitter) and Short-wave Tuner (Receiver) Mark III

Although the BF trench set was to continue in use until almost the time of the armistice in 1918, new apparatus soon was introduced which consisted of a separate transmitter and receiving set. The Wilson transmitter employed a fixed spark gap with high voltage supplied from an induction coil as for the BF trench set, but with a high-speed rotary interrupter instead of the magnetic interrupter. This element of the transmitter meant that the output radio frequency signal was interrupted at an audio frequency set by the rotational speed of the interrupter and as a result had a musical quality which made for easy reception that was not seriously affected by static or other signals in close proximity to the working wavelength of the transmitter.

The Wilson transmitter operated on the same wavelengths as the BF

Wilson transmitter (PRJ)

Short-wave tuner Mark III (PRJ)

trench set but ran at an input power of 120 watts, which produced an improved range of 8 kilometres (5 miles) or more depending on local conditions. This increase in range was also a product of the more efficient use of electrical energy that was inherent to the use of the rotary interrupter. The power supply in this instance was a 26 or 28 volt lead acid battery with a 5 amp-hour capacity.

The companion receiver was the Short-wave Tuner Mark III, which had originally been designed for use with the aircraft-borne Sterling spark transmitter, and had a tuning range of 100 to 600 metres. Like the BF trench set, it used a carborundum crystal for reception with its bias battery but also was able to use the dual crystal Perikon detector. This detector (its development is described in Chapter 4) operated without electrical bias and was a good deal more sensitive than the carborundum detector but not as robust and resistant to battlefield shocks.

With the introduction of tanks in November 1917, wireless became the means by which the new mobile battlefield could be serviced and orders given. The apparatus carried into battle by a tank was the Wilson transmitter and the Short-wave Receiver Mark III. At this stage of development, however, for successful communication the operator had to remove the apparatus from the tank, erect an antenna outside on the ground and set up an earth connection.

Continuous Wave Valve Wireless in the Trenches

By now the French had developed a new hard vacuum valve for military purposes that replaced the fragile and insensitive Audion of De Forest and was known as the R valve. This new equipment was put into operation in 1917 as the German Army retired towards the Hindenburg Line at Vimy Ridge and Arras, culminating in the capture of Passchendaele Ridge.

Despite the success of the spark-based trench sets using crystal detectors, around 1916 the British Army wireless experts at Woolwich began an assessment of new apparatus based on the triode valve. This continuous wave (CW) equipment was capable of operations over a distance of 11 kilometres or more. Operating on relatively short wavelengths, these wireless sets were able to operate on 'low and short' antennas as a result. This latter characteristic would have been extremely popular with signallers used to having their BF trench sets' substantial antenna masts used as aiming points for snipers and concentrated machine-gun and rifle fire.

Despite progress with valves and CW, the spark was not to be entirely displaced when the very compact 'loop set' was introduced as a result of the work at Woolwich. This equipment was designed to be carried by just two men and involved minimal difficulty of installation. The loop set came in two configurations, intended for either forward or rear operation, which were not interchangeable. This is shown in the schematic on page 92.

Forward spark transmitter (PRJ)

Forward valve receiver (PRJ)

In the forward version, the wireless was effectively a hybrid system with the transmitter using a spark gap directly associated with a square loop aerial and having a separate two-valve receiver. The receiver was fed from its own wire aerial with an earth mat below, whereas the transmitter loop aerial was designed to be supported on the 400 mm bayonet used with the standard Lee Enfield No. 1 Mark III* rifle. As the schematic reveals, the forward loop set was quite simple in construction despite being set up in two wooden boxes which would be separated by some distance when in operation.

The apparatus used in the rear location was generally similar to the equipment employed in the forward trenches, with the exception that transmitter and receiver shared a single elevated wire aerial.

Naval Ship's Wireless

From the earliest days of wireless, the Royal Navy had been far more interested in the potential of the new technology than had the British Army.

Even before Marconi demonstrated his equipment in England, the Navy's Captain Henry Jackson had been experimenting with methods to control the movements of the newly developed steam-turbine motor-torpedo boats which could very quickly disappear over the horizon. His experimental wireless equipment, on display at the Royal Naval Communications Museum at Fareham, near Portsmouth, has a clear similarity to that of Marconi. (It has been suggested by some that this apparatus and Tesla's thunderstorm detecting apparatus deny Marconi the title of 'inventor of wireless', an assertion strongly disputed by this author.)

At the start of the First World War, ship's wireless involved a formidable 5 kilowatt rotary spark transmitter. At sea the considerable weight of this apparatus was not a problem, although the disadvantages of spark signal generation remained. The advantages of CW as generated by valves were not lost on the Royal Navy, but because of the distances over which communications needed to be maintained there was less pressure than in the military to jettison the spark system. In addition, with vessels widely separated, mutual interference was far less of a problem than on land, and effective range was considerably more important.

The Quenched Spark
In the spark-generated signal's final manifestation, the Telefunken-developed quenched spark was progressively employed, even by the Marconi Company.

In Australia, where Amalgamated Wireless (Australasia), or AWA, had taken over the local arm of Telefunken at the outbreak of war, the quenched spark system appeared in a ship's wireless system which also featured early valves in a receiver built in a modular system that allowed considerable flexibility. This 'panel set' made it possible for marine owners to specify arrangements most appropriate to their needs.

Wireless in the Air
The British Army remained generally sceptical of the benefits of wireless communications for a considerable period, although its use in airborne spotting of the results of artillery bombardment did find early acceptance and led to the development of trench-based wireless apparatus. The Sterling transmitter, used in conjunction with the Marconi Short Wave Tuner Mark 3 (see page 84), led to the development of ground-to-air communication

AWA panel set receiver and quenched spark transmitter, 1918 (AWA)

that was greatly refined when hard vacuum valves became available in 1917–18.

Before these applications of airborne wireless occurred, in 1915 a young Royal Flying Corps officer, Captain Hugh Dowding, assisted by CE Price, set up wireless apparatus in a Maurice Farman aircraft and became the first person in Great Britain to receive radio telephony from the air. Twenty-six years later, Dowding, by then Air Chief Marshall (and later Lord Dowding), was Britain's principal warrior in the Battle of Britain, in which wireless communications and radar were to prove critical elements in defeating the might of the German Luftwaffe.

Early experimentation with airborne wireless was brought to an abrupt halt when the War Office declared the application of the system to aircraft was impractical. By the time this remarkably shortsighted decision was reversed, valves had made two-way communication a practical proposition. Late in the First World War, airborne valve wireless was in use not only for artillery spotting but for ground-to-air communications. Apparatus typical

Valved wireless as used in World War One aircraft (PRJ)

of the period is illustrated here.

For some time, however, aircraft wireless remained quite primitive, relying on longer wave-lengths which required a trailing aerial to be deployed after take-off. Forgetting to reel in the aerial was an error easily made that could cause damage to the installations on the ground—and to the aircraft—but by 1918 could be seen the genesis of ground-to-air communications and fighter control that would later become critical in aerial combat.

Tank Warfare and Wireless

In the early years of the twentieth century fear of war was common, and a number of writers anticipated events of the Western Front and on the high seas. Particulary notable were two stories published in 1903, a novel by the Irish author Erskine Childers (who much later was branded a traitor and executed by firing squad) and a short story by science fiction writer HG Wells.

Childers' book, *The Riddle of the Sands*, anticipated the expansionary ambitions of Germany and the possibility of an invasion of Great Britain, launched from the Friesian Islands across the North Sea. Such an event never occurred during World War One but much later was contemplated as *Unternehmen Seelöwe* (Operation Sealion), to be launched across the English Channel during World War Two. In the event, Sealion was

Mark V tank, Australian War Memorial (PRJ)

abandoned when the initial phase of aerial attack failed in the Battle of Britain.

HG Wells' short story, 'The Land Ironclads', came very close to specifying the weapon that would end the deadlock of trench warefare, the tank. His description of the land ironclad, a name referencing the ironclad battleships used in the American Civil War, as 'something between a big blockhouse and a giant's dish-cover' comes quite close to what was to eventuate in the trapezoidal form of Great Britain's first tanks. As the tank was later developed, this description became ever closer to the fact, as the illustration of the Mark V tank of 1918 reveals.

In reality, it was not until the outbreak of war that a serving military engineering officer, Lieutenant-Colonel Swinton, suggested attaching armour plating to a tracked vehicle developed in America for use on rough terrain: the Holt tractor. Concurrently a French officer, Colonel Estienne, had the same inspiration and the development of this new device proceeded in parallel in Great Britain and France, although on somewhat different physical lines. There has to be a reasonable suspicion that the designs for Lancelin De Mole's travelling caterpillar fort had been examined at the War Department and the idea taken up. Although never overtly acknowledged, the belated award to De Mole of money sufficient to cover the cost of drawing and model-building suggests a tacit acknowledgement of the precedence and usefulness of his work.

After considerable experimental work had been undertaken—to yet another chorus of military scepticism—the idea of the armoured land vehicle was strongly supported by the Secretary for War, Lloyd George, and the First Lord of the Admiralty, Winston Churchill. Much of the early development work was carried out under the auspices of the Royal Navy, associated with the notion of a new armoured vehicle which would resemble a Dreadnought on land—a 'Landship'. As a disguise for this highly secret development, the name 'Tank' (as in water tank) was adopted and has remained attached to this machine of war ever since. The necessity for this new device to surmount the physical barrier of barbed wire and enemy trenches resulted in the characteristic, rather peculiar trapezoidal shape of British tanks in the First World War.

In September 1916, on the Somme battlefield, 24 first-generation tanks crossed enemy trenches with relative ease but in other respects their success was limited, resulting from the failure of higher command to understand how this new tool of warfare was best employed and followed up with infantry consolidation. Again at Cambrai in 1917, where conditions were generally quite suitable for tank operations, an initial major success was marred by a failure to capitalise on the gains achieved. The assault by 400 British tanks tore a huge gap in the German front line and tanks penetrated to a distance of nearly 6 miles (9.6 kilometres) along a 7 mile (11.25 kilometre) stretch. (This can be compared with infantry advances that resulted in territorial gains measured in metres, and where the level of casualties frequently was measured in many thousands.) When these gains were not followed by appropriate infantry consolidation, the German counter-attack saw the territory lost.

A fully successful assault with tanks had to wait until July 1918 at the French town of Hamel, and the involvement of Australia's General Monash, before the correct tactics were applied. This was a fully coordinated attack involving tanks, infantry, artillery and the assistance of aerial observation and spotting by aircraft of the newly formed Royal Air Force (which replaced the Royal Flying Corps in April of 1918). The result was a crushing attack on the German front line and a reverse that was described by Germany's General Ludendorff as the 'Black Day' of the German Army. This loss can be seen as marking the beginning of the end for Germany.

The new armoured vehicle was relatively mobile and accordingly hard to maintain contact with. For this reason, it was realised early on that wireless was an essential aid to command and control of the fleet of tanks as

Tank signals and apparatus (RAS)

they moved ahead of the advancing infantry. While the form of wireless then in use was quite primitive, and to use it required the tank to come to a halt so that a mast for the aerial could be erected, the way ahead was firmly established. Ironically, the importance of wireless in this context was most clearly perceived by the defeated German Army, an insight put to devastating effect in later development of the tank.

From their operators' perspective, the first-generation tank of World War One must have been a truly awful device in which to fight. Lacking anything approaching a sprung suspension, drivers and gunners were exposed to violent impacts from rough terrain against which not even a leather helmet could protect them. Engine noise was extreme and exhaust gases were not properly evacuated, so that the interior soon filled with a dense cloud of fumes mixed with cordite from the guns.

As the accompanying photograph reveals, the installation of wireless equipment was also something of an afterthought and the signaller was forced to operate the best way he could. When the need to transmit a message arose, he was forced to clamber out of the tank, potentially under enemy fire, and erect an antenna mast before making contact. Not at all an attractive proposition—but it led the way to later developments in which tank-to-tank communication was made possible from within the protective hull.

World War One Wireless in Retrospect

This was a war in which both sides expected to achieve an early victory and the commanders frequently rode off to battle on horseback as if on a modern crusade. Stalemate on the Western Front was the inevitable result of this lack of military imagination and foresight and it was only with the invention of the tank and the return of battlefield mobility that this impasse was overcome.

Wireless had been useful in the initial months of the war as the rapid movement of the attacking German forces pressed the Allies back to the Marne. However, once the war became static, this new device was set aside in favour of the telephone. In the trenches, wired communications initially provided the necessary links from the front line to commanders at the rear, but the telephone suffered from two problems that brought wireless back into consideration. Electronic interception was possible and the German Army managed to obtain much critical information that resulted in the loss of many lives, but what really spelt the end of wired communications was shellfire, which tore up the wire links and created massive casualties among the signallers sent to repair the damage.

By the end of the war, wireless had become firmly established as a reliable means of conveying messages, even where this involved the relatively short distances that were required to be crossed behind the static trench system. In the more open and mobile warfare that was waged in Palestine and Mesopotamia, and later in the Dunsterforce advance into Persia, mobile wireless stations became the essential means of sending messages over distances up to 160 kilometres (100 miles). The Marconi apparatus had at last convinced the Army that here was a system that was now indispensible—and since its first exposure in 1896, any doubts had been dispelled.

World War One was a truly awful conflict in which far greater numbers were killed than in any war before or since. If there could be seen to be any benefit from all this slaughter, it at least enforced a different attitude and approach in the conduct of the next world war, twenty years later. In particular, the use of tanks as the spearhead of attacks, rather than fully exposed infantry, helped to preserve many that under the doctrine of World War One would not have survived.

The other clear benefit was the impact on the science and technology of wireless, in particular its application to civilian uses in radio broadcasting, impossible without the services of the newly developed thermionic valve. This device would be the key element of radio communications for the best part of the next forty years.

4 Technological Change: From Spark to Valve

For the first twenty years or so of wireless telegraphy, the source of radio frequency energy had been the spark which, since time immemorial, nature had presented to humanity as bolts of uncontrollable Jovian lightning. Almost from the beginning, though, scientists had recognised that the radio frequency energy contained in a spark was an untidy form of electro-magnetic noise rather than a pure electrical wave. Moreover, when pressed into service in this new method of communication, the spark was very energy inefficient. Early efforts by Lodge and then Marconi achieved some progress in improving energy efficiency by limiting the range of frequencies over which the band of spark-generated radio frequency energy was distributed, but the spark was intrinsically incapable of producing the single continuous sine wave that very soon became the goal of radio engineers.

Initially this process of increasing selectivity was referred to by the interesting word 'syntony', but over time came to be known as 'tuning'. Where early spark transmission consisted of a burst of radio frequency energy which quickly faded away in what was referred to as a logarithmic decrement curve or 'log dec', it was appreciated that a continuous sine wave of radio frequency energy would make tuning more effective and be more energy efficient. At quite an early stage, Canadian scientist Professor Reginald Fessenden was able to demonstrate the production of continuous-wave radio frequency energy with the assistance of an alternating current generator. This was operated at a frequency far above the conventional 60 hertz of mains alternating current electricity but produced very low power output as compared with spark-generated energy.

As the science of wireless was progressively developed, it was realised that the frequency at which a radio wave would occur could be determined by changing inductance and capacitance in the transmitter and receiver circuits associated with the aerial and earth. This expansion of scientific knowledge found its first and most enduring expression in a device invented in 1904 by CS Franklin, an officer of the Marconi Company, when he was in Russia commissioning one of the early Marconi wireless stations. The multiple tuner was patented in 1907 and became a standard element in a Marconi wireless station for many years. Over a hundred years later it remains at the heart of modern tuning devices and methods.

Multiple tuner, 1904 (MAR)

Spark and Continuous-wave Radio Frequency
An alternative to the energy inefficient system of spark wireless communications, the radio frequency alternator was developed for Fessenden by Ernst Alexanderson of the US electrical engineering company General Electric (GE), under the supervision of its engineering head, CP Steinmetz. In August 1906, one of three high-frequency alternators was delivered to Fessendon by GE. At this stage, the frequency was limited to 50 kilohertz, with a very low power output compared to the rotary spark generator then in common use.

Despite these limitations, the operational attributes of the early Alexanderson alternator pointed the way ahead to a fully continuous wave system. Later, in 1924, the by then obsolete Alexanderson alternator transmitter was used in a high-power long-wave station that operated in the vicinity of 17 kilohertz. This station, at Grimeton in Sweden, was able to provide a telegraph service between the United States and Sweden, an economical alternative to the undersea cable service that had been available since 1866.

The Grimeton VLF radio station remains as the sole continuous-wave alternator transmitter able to produce a pure sine wave. It is still used on special occasions, such as Alexanderson Day, to transmit Morse messages on 17.2 kilohertz. Its call sign is SAQ. In 2004 it was declared a World Cultural Heritage site by UNESCO.

During the years leading up to the First World War, despite trenchant support for the superior qualities of spark for telegraphic communications, Marconi worked on improving the characteristics of the spark-generated radio frequency signal. Initially this was achieved by employing a rotating spark gap, using both asynchronous and synchronous spark generation. This method was employed in the pack set of 1913 and in the cart and motor vehicle-mounted wireless stations that went to war. Later the quenched spark gap developed by Telefunken was employed, particularly on seagoing vessels.

In the trenches, in the first two years of the war, spark was the only method of producing a wireless signal from anything resembling truly portable apparatus that could be carried by the infantry. The first experiments were undertaken with a transmitter referred to as the Sterling, which was used in aircraft to assist in spotting the fall of shot for the artillery. The compact 30 watt transmitter was able to send Morse-coded messages to a ground-based receiver operated by the artillery batteries. This receiver was a form of crystal set in which carborundum was used to convert the spark signal to an audible form. No receiving capability was provided to the pilot of the spotting aircraft—in any case, the high audio noise level of an open cockpit and intense electrical noise would have made such a facility redundant.

Sterling transmitter (MAR)   Marconi crystal set (PRJ)

The successful use of the transmitter and crystal set encouraged previously sceptical Army authorities to request the development of a compact short-range transmitter and receiver package, similar to the Marconi pack set of 1913 (see Chapter 2). Commonly known as the BF, this wireless provided a very effective substitute for the damage-prone wired telephone system. The BF trench set was simple in operation and not intended to tax the mind of the harassed signaller under fire from the other side of No-Man's-Land. Its simplicity is made clear by the accompanying schematic.

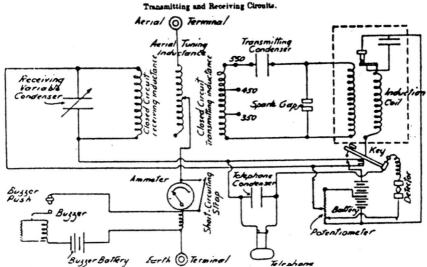

Schematic for the BF trench set (LM). Larger version p. 320.

Concurrently with the spark developments on land, efforts were made to improve the quality of the spark in the trans-Atlantic wireless station established at Clifden in the west of Ireland in 1907. When this station became obsolete, the rotary spark system was also incorporated in the station built in 1912 at Wynfawr in North Wales, close to Caernarvon. Through increasing the number of spark electrodes and the number of rotating spark plates, Marconi believed he was able to create radio energy with the characteristics of the continuous wave, but even in its ultimate form as the 'timed spark' system this was not true. Nonetheless, this form of spark transmission did produce a very satisfactory audio signal for wireless operators when receiving Morse code messages, and the same approach was taken in the progressive refinement of marine wireless apparatus during the later part of World War One. The fundamental problems of spark-generated radio frequency energy remained, however, and efforts to improve its quality continued.

Wireless apparatus on seagoing vessels was increasingly provided with the Telefunken-style quenched spark generator, a device that had the beneficial characteristic of generating a high-frequency tone (the 'singing spark') that was able to be read through intense interference and in the presence of competing spark-generated signals.

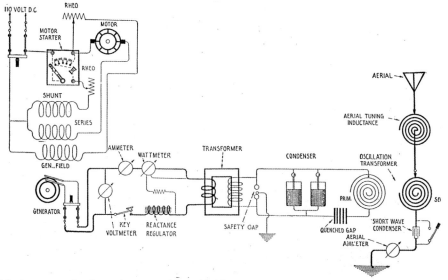

Marine quenched spark transmitter, 1916 (INT). Larger version p. 320.

The spark was eventually made obsolete by a device developed by the senior scientific advisor to the Marconi Company, Professor Fleming. Suffering from a degree of deafness, in 1904 he had searched for a new means to detect radio frequency energy that would not rely on the hearing of an audio signal as provided by the coherer and, later, the magnetic detector.

Detecting Radio Frequency Energy

Heinrich Hertz's experiments in the 1880s to determine the existence of radio frequency energy involved a single loop of wire and a minute adjustable spark gap that necessitated the use of a magnifying glass to identify the occurrence of a spark. Clearly impractical as a means of detecting radio frequency energy other than in experimental mode, a better device was needed once the idea of communications was developed. The first such detector was the coherer, a delicate and relatively insensitive device developed by the French scientist Edouard Branly, and based on the earlier discoveries of the Italian physicist Temistocle Calzecchi-Onesti. Marconi and other early wireless experimenters employed the coherer in the first generation of wireless telegraphic apparatus.

Following the successful trans-Atlantic experiment of 1901, it was clear that a better method of detection was required, and Marconi set to work to find a solution. Based on the earlier experiments of Ernest Rutherford, Marconi created a device which, until the First World War, was the main detector used in his company's wireless stations—the magnetic detector or 'Maggie'.

The magnetic detector (MAR)

The Maggie incorporated a multi-stranded wire rope which ran around two circular spools which were driven by a wind-up clockwork mechanism. The wire rope ran through a small solenoid with two windings, one of which was fed from the aerial through the primary coil to earth while the other was connected to a pair of earphones. By a physical process that remains obscure even today, through the movement of the wire rope through the solenoid a pulse of radio frequency energy was converted into audio. The magnetic properties of iron wire have been suggested as the basis of this conversion, and the audio pulse heard in the earphones is thought to be generated by the radio frequency pulse that passes through the moving wire rope. This is then associated with a process of hysteresis in an area of local magnetisation in the wire rope produced by the radio frequency energy pulse. The audio pulse is heard when, almost immediately, de-magnetisation of the local area of the rope occurs.

The magnetic detector was a highly reliable detector of radio frequency energy, although somewhat insensitive compared to the valve, the device that eventually replaced it. Its major disadvantage was that it required constant attention, the clockwork mechanism needing to be wound up periodically to maintain the movement of the wire rope.

The Valve as Detector and Amplifier

Prior to the successful 1901 experiment which saw Great Britain joined to Newfoundland by Morse code, the Marconi Company had referred serious problems of back electromotive force (EMF) in the high-voltage power generator to its newly appointed scientific advisor, Professor Fleming, who had been able to overcome significant problems with the proposed spark transmitter. Of far greater significance were Fleming's experiments with the evacuated light bulb, invention of the US experimenter Thomas Edison.

Fleming's increasing deafness induced him to look for a method of radio frequency energy detection that would not require an audio signal to be heard. The Edison evacuated bulbs had indicated that impressed alternating current would be rectified: this was the basis of the so-called 'Edison effect'. Fleming wondered if this ability to rectify an alternating current would also apply at radio frequencies—and experimentation showed that indeed this was the case.

The resultant device, now called the thermionic valve or diode (tube, in the United States) and still rather insensitive, was in essence a conventional light bulb, but in addition to the electric element that glowed white hot a second electrode was sealed inside the evacuated glass bulb. This vital invention occurred in 1904 and subsequently the Marconi Company supplied the valve as a detector of radio frequency energy up to the beginning of World War One.

Within a very short space of time the diode was to be transformed by the enterprising young US inventor Lee De Forest. Using a number of diodes manufactured for him by the HW McCandless electric light bulb factory in New York and which derived directly from Fleming's invention, De Forest began to experiment with supplementary internal connections, ostensibly related to earlier work he had undertaken with gas flame detectors. From this came the introduction of the 'gridiron', subsequently referred to as the grid, which allowed control of the stream of electrons between the heated cathode wire of the diode and the anode plate.

Fleming valve (PRJ)

These early three-element tubes, named Audions, had a relatively low vacuum and although rather temperamental proved quite useful as amplifying detectors. Further research, notably that of Dr Irving Langmuir of GE in the period immediately before the First World War, revealed that increasing the vacuum in a tube so that it was deemed to be 'hard' converted the triode into a stable and reliable amplifier of audio and radio frequency energy. This discovery was soon exploited, as the spark transmitters and crystal receivers of 1914 were subject to mutual interference from other operators on the battlefield and the need for much 'sharper' tuning became critical.

By that time, the temperamental Audion had been tamed by the imposition of scientific examination in an industrial environment and, with a hardened vacuum, could be made to oscillate via positive feedback. This would become the heart of wireless reception and transmission in the years to come, ousting the crystal detector in receivers in the process.

Crystal or Solid-state Signal Detectors

Up until the early 1920s, another useful radio frequency detector was available in the form of mineral crystals operating as radio frequency rectifiers. The property of rectification had been discovered by Professor Braun long before the experimental work of Marconi, but became relevant to wireless telegraphy as a result of the development work of the US radio pioneer Greenleaf Whittier Pickard in 1906. Working with galena (also known as silver lead), he discovered that with the application of a fine wire

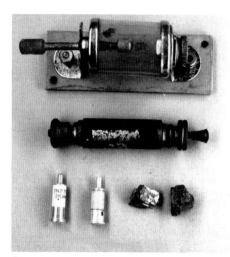

Clockwise from the top: cat's whisker, Perikon detector, samples of galena and two World War Two radar diodes (PRJ)

to this mineral, radio signals could be detected. This device became known as the 'cat's whisker' detector and not only survived the 1914–18 war but served the first generation of broadcast listeners after 1922. Later still, in 1947, the crystal form of germanium was to make a reappearance as the basis of the first solid-state amplifier, the transistor.

Pickard later discovered that a very effective rectifier could be made from the conjunction of two minerals, zincite and chalcopyrite, held together by spring force in a tubular holder. He called this device the Perikon detector ('Perikon' being an acronym for 'perfect Pickard contact'). As it was significantly more stable than the cat's whisker, it became very popular.

At about this time, General Henry Harrison Chase Dunwoody, chief signal officer of the US Army, discovered that carborundum (silicon carbide) was also capable of rectifying radio frequency energy although, for best operation, a bias voltage needed to be applied to the device. Despite this limitation and its somewhat reduced sensitivity compared to the cat's whisker and the Perikon detector, the carborundum detector was much less affected by severe vibration and thus became very popular where extreme motion was likely to be encountered, notably on naval vessels subject to heavy gunfire at sea.

Valves in the Trenches

The stalemate on the Western Front, and the known advantages of short-range wireless links, drove demands for better apparatus. The spark sets of 1914 and 1915 were about to be replaced with apparatus incorporating the new hard vacuum valve. The valve at first was provided only in receiving apparatus, to increase sensitivity and range over the unamplified audio output of the carborundum crystal set, and the transmitter remained unchanged.

The upgraded trench wireless supplied in 1916 consisted of two sets of transmission and reception equipment, with valves now used in the receivers. One set was designed for the forward trench location, the other set for reception and transmission at the rear.

These two diagrams show the connections between transmitter and receiver for both the forward and rear wireless apparatus as well as the aerial arrangements. Both sets were designed to work on wavelengths of either 65 or 80 metres. In the case of the rear wireless set, the transmitter and receiver used the same wire aerial which for 65 metres wavelength operation was approximately 15 metres long and for the 80 metre wavelength approximately 18 metres long. Both forward and rear transmitters produced about 20 watts of RF power and had a relatively short range of between 1.5–2 kilometres.

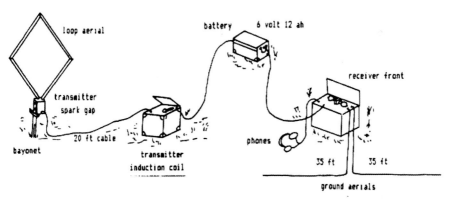

Forward spark transmitter–valve receiver and aerials (LM). Larger version p. 321.

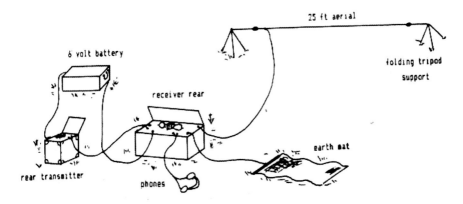

Rear spark transmitter and valve receiver with aerial (LM). Larger version p. 321.

In the schematic below of the forward wireless equipment, note that the receiver is to be read from right to left, rather than the left to right reading followed today. The front end of the receiver operates with a positive feedback loop to produce continuous oscillations that are mixed with the incoming spark signal from the rear station. From the mixing of these two signals, which are arranged to be almost coincident in frequency, an audio frequency signal is generated by the heterodyne effect (today referred to as 'direct conversion'). The local oscillations generated in the first receiver valve were set up to be close to the input signal with a separation of perhaps 1500 Hz. The resultant heterodyne or 'beat' frequency audio signal at 1500 Hz, achieved by mixing the two signals, was then amplified in a second audio valve.

This method of reception is still encountered, mostly in simple receivers for both CW and single sideband reception, particularly in amateur radio operations. It was but a short step away from the superheterodyne process developed by EH Armstrong that is discussed in the next section.

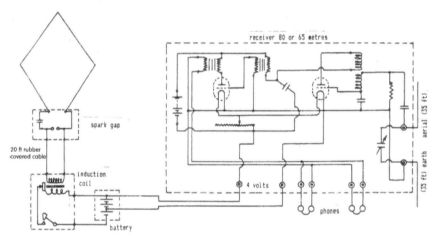

Schematic of forward loop set (LM). Larger version p. 321.

Regeneration and the Superheterodyne

It was not until the decipherment of the infamous Zimmerman Telegram of February 1917, a diplomatic proposal from Germany that Mexico join Germany in making war against the United States, that the United States finally decided to join the Triple Entente and send troops to France. Apart from the supply of guns and shells backed by US industrial muscle, and the size of the infantry forces deployed, the US Signal Corps contributed the services of a young officer who was to have a remarkable long-term impact on the future of wireless communications. Edwin Howard Armstrong, as a very young man, in 1912 invented the regenerative receiver and later, during his army career, the superheterodyne principle of radio reception. Much later, Armstrong went on to invent frequency modulation, a radio frequency static-resistant system of broadcasting and reception.

While still an undergraduate engineering student at Columbia University, in 1913 Armstrong had 'played' with De Forest's Audion and had discovered that by feeding back some of the amplified energy in the triode, it could be made enormously more sensitive when used in a radio receiver. Partial feedback, known later as 'reaction' (or, more correctly, 'regeneration'), was a staple element of early wireless receivers well into the mid-1920s. If further energy was fed back to the input grid of the valve, continuous oscillations occurred and created the long-sought 'continuous wave'. This principle was soon applied to World War One military transmitters (see Chapter 4). The regenerative detector receiver was eventually superseded by another Armstrong invention, the superheterodyne receiver.

Captain EH Armstrong (*WW*)

Regeneration was the subject of fierce litigation by Lee De Forest (who brought so many actions against other inventors that today he might well be classed as a vexatious litigant), in which he managed to persuade an American court that he had invented it prior to the youthful Armstrong's discovery. Reference to material associated with this case suggests that the judge involved was technically ignorant. He was led to believe that producing a feedback 'howl', as noted in a De Forest notebook, could be equated with discovering the subtler application of controlled feedback that is fundamental to the operation of a useful regenerative receiver. Undiscouraged by these judicial proceedings, which ran on interminably, Armstrong joined the US Army Signal Corps and continued to carry out experiments with the triode-based method of radio reception.

While working on problems with detecting low-level radio signals in the trenches, Armstrong developed the superheterodyne ('superhet'), in which a local oscillation is mixed with a received radio signal to produce a difference signal at another frequency above audio frequency. The new signal could be arranged to occur at a fixed frequency well below the received signal, and could also be amplified by a highly selective chain of amplifiers at the so-called intermediate frequency. His patent for this procedure was overturned in 1929 by a patent granted to the French experimenter Lucien Levy, but by that time the superheterodyne process had become universal. It remains the basis of virtually all radio receiving devices used today. Armstrong's last great development, frequency modulated broadcasting, is considered in Chapter 5 in the context of civilian and military use.

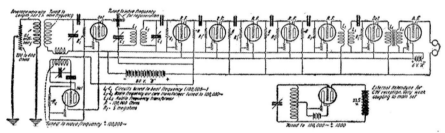

Armstrong superheterodyne receiver—first amateur trans-Atlantic contact (*WW*). Larger version p. 322.

This diagram of the Armstrong receiver used in Scotland in 1921 by Paul Godley to detect signals sent by radio amateurs in the United States will be seen by the experienced to have some peculiar features compared to later superhet receivers. In particular, the chain of amplifiers at the intermediate frequency (IF) is resistance-capacity coupled, whereas later superhets employ inductive coupling via intermediate frequency tuned circuits. No doubt this peculiarity was related to the characteristics of the triodes available at that time. The early triodes had relatively low gain and there was also the need to avoid uncontrollable positive feedback—a problem that in the future would be overcome by neutralisation of the IF stages. Later still, multi-element valves would eliminate problems of internal capacity and the tendency to self-oscillation.

PROJECT 1: Wilson Transmitter Replica

The British Expeditionary Force that crossed the Channel to France in 1914, the First Hundred Thousand, had been sent to participate with French forces in a war that it was assumed would feature cavalry-based mobility. For a while, the new battlefield was indeed characterised by rapid movement, but from the banks of the River Marne the stasis of the Western Front supervened.

Abandoned with the onset of the static war of attrition was the early wireless apparatus that the Expeditionary Force had brought with it. The horse-drawn cart wireless and the pack sets intended to be transported by packhorses were now found to be of little tactical value. What was needed was a small, portable wireless that could operate out of the trenches. It would need only a relatively short range, and would have to be light enough to be transported by a small group of soldiers. Although somewhat underpowered the BF set (British Field set) met these requirements and was soon introduced with considerable success.

Sterling transmitter (PRJ)

As the Western Front battlefield stabilised and expanded laterally on either side of No-Man's-Land, the increasing range of enemy shellfire meant that battalion headquarters needed to be located progressively further from the front line. Now a somewhat more powerful wireless set was required and from this came a new device, the Wilson transmitter, which entered service during 1916. Unlike the single-unit BF set, transmitter and receiver were now separated. This arrangement presumably derived from the earlier success of the airborne Sterling transmitter and separate Marconi crystal receiver used for artillery target spotting.

The Wilson transmitter and its companion crystal receiver provided a generally reliable means of communication over a distance of several kilometres. Boxed in a very similar manner to the BF trench set, the Wilson transmitter employed a motor-driven interrupter to create the spark and was thus credited with having a 'musical' and readily distinguished sound in the earphones that was claimed to overcome interference from other spark stations on both sides of No-Man's-Land. While rather less ragged and raucous than the output of the BF trench set, the term 'musical' was more a matter of optimism than reality.

BF trench set in operation (RAS)

Short-wave tuner (PRJ)

Construction

The general arrangements of the Wilson transmitter appear in the schematic opposite, and the simplicity of the system is readily apparent. However, construction of the hardware is significantly labour intensive, although the end result is visually satisfying.

The timber box case of the Wilson transmitter displayed at the Royal Signals Museum at Blandford Forum in Great Britain is 380 x 300 x 275 mm (15 x 12 x 11 inches), the basic parameters for a container suitable for the replica. To the author, a long-term hoarder of interesting historical electrical artefacts, one item immediately suggested itself as potentially suitable for this role. This was a portable, very high-voltage medical device from the 1920s which was used for epidermal stimulation. The high

PROJECT 1: Wilson Transmitter Replica

voltage and associated electrical discharge was generated by means of a magnetic self-actuated interrupter and high-voltage induction coil, and all the components were contained in a box of almost exactly the dimensions required for the replica.

Given the alternatives of constructing a new box from raw materials or converting an existing box made of nicely aged English oak, there was very little hesitation in deciding to follow the latter course. The remains of the medical apparatus were set aside for potential projects of the future, and the front of the box was cut off and remounted on hinges, allowing access to a lower compartment for the spark gap, rotary interrupter and induction coil assembly. When stripped down, this medical apparatus also provided a set of four screw-threaded terminals which could be relocated on the top panel of the replica transmitter.

Wilson transmitter at Royal Signals Museum, Blandford Forum (SMB)

Located under the top panel, in the base of the box, and visible when the front panel is opened up and laid flat, are the adjustable spark gap and the local

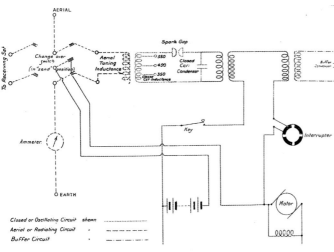

Schematic for Wilson transmitter (LM). Larger version p. 323.

oscillator coil. Loosely coupled to this coil is the aerial coupling coil, together with its taps, connected to the rotary aerial coupling switch on the panel above. At the rear of this space is the induction coil (a 12 volt DC motor vehicle ignition coil from an automotive wrecker's yard), and in front of it the motor-driven interrupter in its metal cradle.

In replicating the Marconi multiple tuner (described in the author's earlier book *In Marconi's Footsteps: Early Radio*), a panel of black Formica switchboard material was left over. This was sufficient to construct the top panel of the Wilson transmitter. Formica is an extremely hard and dense material, necessitating the use of a plunger-type jigsaw with a metal cutting blade. Cutting out the opening for the output meter required the use of an electric drill-driven rotary cutter—which needed sharpening more than once during the process.

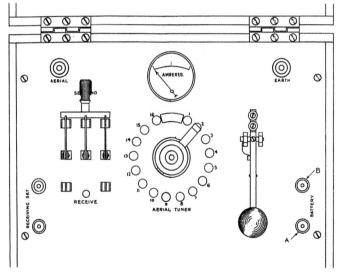

Top panel of Wilson transmitter (LM)

The multiple stud aerial matching switch in the centre of the top panel requires 18 brass studs. As no comparable item is available commercially, this demanded a protracted process of brass rod cutting on the home lathe. Firstly a 3/8 inch diameter rod was reduced in diameter to create a ¼ inch diameter shaft that was threaded with a ¼ inch die; this stud was then cut away from the main section of the rod and reversed in the lathe to allow the top of the stud to be faced and finished. The total length of the shaft was ¾ inch and the thread was cut to 1/8 inch below the intended stud head. Although a seemingly trivial task, achieving a common depth from the face of the lathe on all 18 studs was essential to allow the sweep contact of the rotary switch to move smoothly from stud to stud.

PROJECT 1: Wilson Transmitter Replica

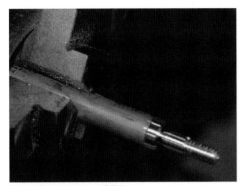

Stud fabrication (PRJ)

Following the stud cutting operation, holes were drilled in the Formica and the studs installed as shown in the top panel schematic. This required ¼ inch brass nuts—which initially proved impossible to find in the usual commercial hardware stores. As a stopgap, 18 circular threaded stud holders were cut out of a pre-drilled and tapped brass rod and applied to the grinding wheel to create opposed flats able to take a spanner. Later, ¼ inch brass nuts were obtained from a specialist company, and the fabricated stud holders were retained as spacers for wire tags.

The triple-pole, triple-throw surface-mounted switch presented a further problem. In the end, the essential components were sourced from two modern double-pole switches. The spring-loaded contacts were removed, together with the hinged centre contact arms, and assembled on a new hardwood base which was mounted on top of the Formica base board—a slight departure from the original, where the contacts were fixed directly to the top panel, but the simplification of construction more than justified this liberty. A brass actuating knob to sit at the centre of the new top bar connecting the three sets of switch arms required a further visit to the shed and the home lathe.

Directly associated with the rotary interrupter used in the Wilson transmitter is the fixed, self-cooling spark gap unit, which again required lathe fabrication, this time from aluminium rod obtained from a specialist metals supplier. As the illustrations here and on the next page show, grooved cylindrical spark knobs were fabricated with threaded fixing and adjusting rods which in turn were carried on vertical

Spark gap component parts

pillars, also made of aluminium. Fixed to a plate of Formica, this assembly required the construction of two setting and locking knobs, in this instance made from black Delrin rod left over from another project. The outer knob was cut and drilled, then tapped to match ¼ inch threaded steel rod, and then a second, larger knob was constructed. This latter knob is used to lock the threaded rod that sets the spacing of the two spark knobs. The outer knob was fixed to the adjusting rod with a threaded pin, cut down from a 1/8 inch round-headed bolt.

Spark gap (PRJ)

The Morse key, home-constructed by a radio amateur many years ago, was obtained from a sale at a field day sometime in the past. As it is close in size and design to the key used on the original transmitter, it was pressed into renewed life here, thereby saving considerable time and effort. Given the amount of labour next involved in constructing the rotary interrupter unit, this was certainly beneficial.

A particular difference between this transmitter and the BF trench set is the method of activating the induction coil which in turn generates the high-frequency spark. For the replica a DC motor purchased as a spare some years ago was pressed into use. It should be noted that creating this interrupter relied on the availability of relevant parts from the world of the automobile—which still employs components introduced a century ago, in particular the petrol engine Kettering ignition system, invented in 1911, which provided the necessary induction coil. The more recently developed alternator generator system also provided essential parts.

At one stage, it had been hoped to have the rotating bronze interrupter block cut with a commercial milling machine provided by a fellow radio amateur but in the end resort to

Interrupter rotor block (PRJ)

traditional methods became necessary. A fine-toothed hacksaw, a 6 mm wide square file and strips from a sheet of Ebonite were employed in creating the rotating element of the interrupter.

Having set out the slots in the interrupter rotor block at 60 degrees, these were cut out, firstly using the hacksaw in a series of parallel cuts and then enlarged with the file to six parallel trenches. Into these six trenches, strips of ebonite were fixed with the two-part epoxy glue, Araldite, and later filed back and emery-papered to a smooth finish.

The rotor block was then fixed back to the DC motor shaft with a threaded stud, set into a small trench filed in the shaft of the motor. The interrupter brushes were then made from a Lucas alternator brush assembly obtained from the old stock of an auto-electrician. The brush blocks required slight trimming to fit the width of the Ebonite strips so as to avoid shorting across the width of adjoining sections of the bronze rotor block. As can be seen from the illustration, the Lucas unit was bolted back to a small folded metal carrier, which in turn was carried on a threaded rod bolted back to the cradle for the motor. In turn, this carrier was spring-loaded to ensure brush contact with the bronze rotor.

The interrupter motor is carried on a metal cradle which also provides a place to attach the rotor brushes, as the accompanying illustrations show. The cradle is supported on four rubber grommets and is bolted through these items to the underside of the wooden transmitter box.

The final elements in construction were the internal and external tuning coils and the internal fixed capacitance. This latter element was

Interrupter rotor and DC motor in cradle (PRJ)   Rotor and brushes (PRJ)

constructed of alternating glass and galvanised sheet mild steel plates and, by measurement, produced a capacitance of 580 picofarads. This capacitor is connected to the internal coil which consists of 12 turns of 1.5 mm copper wire (obtained from the local metal recycling depot) wound onto a 75 mm diameter cardboard former and spaced apart over a distance of 100 mm, then fixed in position with rapid-setting two-part epoxy glue. To hold the windings in position while applying the glue, strips of masking tape were run down the windings to the cardboard former. The internal coil was then finished with a coating of dark coloured varnish to hold the windings steady.

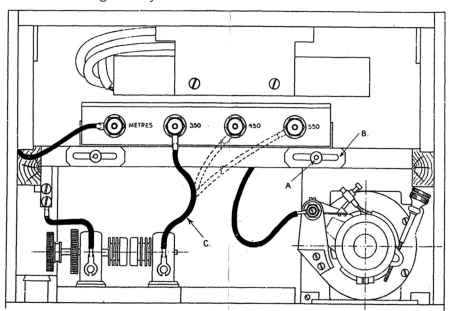

Wilson transmitter with front panel open (LM)

With a certain amount of trial and error, assisted by a solid state dip meter, it was possible to produce an internal network that resonated at 1.8 megahertz by adjusting the number of plates in the fixed capacitor. This frequency is part of the radio amateur allocation in Australia and it is hoped it may be available for the 100th anniversary of the introduction of the Wilson spark transmitter in 1915.

Although the sectional elevation of the Wilson transmitter from the operator's handbook suggests that the tuning inductances were probably wound on spider arms set within square enclosures, using conventional cylindrical coil formers seemed a reasonable concession to absolute

PROJECT 1: Wilson Transmitter Replica

accuracy. This is particularly the case as this part of the transmitter is entirely concealed below the Formica top panel and can only be accessed by unscrewing the front timber panel.

What is also apparent from this diagram is the ability to change the relation of the tapped external antenna coil with the fixed coil in the internal spark circuit. In the original transmitter, this appears to have been achieved by sliding the internal coil box laterally below the external coil box. In the replica this same capability is achieved using cylindrical coil formers. The external diameter of the outer coil was selected at 88 mm. This allows the internal spark circuit coil to slide into the external coil, which allows the necessary loose coupling to be achieved that in turn ensures that reasonably sharp tuning can be obtained—although it has to be said that use of the word 'sharp' tends to be extremely optimistic in the context of a spark-generated signal The more accurate description, historically, is of a signal that is as 'broad as a barn door'. The two cylindrical coils constructed for the replica appear in the accompanying illustration.

Internal spark coil plus antenna coil (PRJ)　　　　Impedance matching (PRJ)

The other critical element of the internal spark circuit is the fixed capacitor (or condenser, as it was then called). This was made up with metal plates separated from each other with glass plates, 100 x 150 mm, and achieved a capacitance of 580 picofarads, measured with a digital meter. The metal plates, as can be seen from the illustration, were finished with rounded corners to minimise the potential for high voltage sparkover, and measured 85 x 150 mm. Three metal plates were interleaved with two other metal plates and separated by the glass plates, with tags soldered to the metal surface and protruding at opposite ends of the capacitor and joined together and connected to the terminals of the spark gap and the internal coil.

The relationship of the internal spark circuit components is shown in the accompanying illustrations with the top and side panels removed during the construction phase. Also visible is the induction coil.

It was decided to employ another device from contemporary automobile electrical arrangements, an alternator filter which, internally, consists of a high frequency choke and a capacitor to bypass the radio frequency energy to earth, to replace the single capacitor of the original. All these elements must be wired together before the top panel and associated antenna matching coil can be inserted into the vacant space in the top of the replica. The internal coil is inserted into the end of the antenna coil at this stage, its final position being adjusted by hand from below before the front panel is fixed into position.

Fixed capacitance (PRJ)

Spark gap, capacitor, induction coil (PRJ)

Internal spark circuit components (PRJ)

PROJECT 1: Wilson Transmitter Replica

Associated with the tuning coils of the replica is an antenna current meter, which in the Wilson transmitter was a moving iron meter that was directly sensitive to radio frequency energy. Such a component was not commercially available but, courtesy of an eminent member of the Historic Radio Society of Australia, a 5 milliamp moving coil meter was provided. This meter had a metal shell which only required the removal of black paint to produce an authentic appearance when projected through the front panel of the replica. An appropriate shunt resistor, together with a modern diode and capacitor, allowed this meter to provide the indication of maximum antenna current that can be achieved when the rotary switch is rotated to the correct stud and a length of aerial wire is run out. For the frequency applicable in this replica, 1.8 megahertz, a 12 metre length of wire, along with a good earth rod or ground-mat of copper wires, is adequate.

Interior of the Wilson transmitter replica showing antenna matching coil (PRJ)

The Wilson transmitter was operated in conjunction with a Marconi carborundum or Perikon crystal set. Changeover from transmit to receive was achieved with the three-pole change-over switch, fabricated from two sets of modern school laboratory switches. The lower pair of poles provides a contact from the crystal set aerial and earth to the external aerial and earth connections. The upper two contacts provide connection from the transmitter to the external aerial and earth; in addition to these poles, the centre pole activates the interrupter motor in the transmit position. The original operator's handbook specifically prohibits using the Morse key unless the interrupter motor is running, to prevent excessive current being passed through the primary of the induction coil.

Finally, an alternator filter was employed in the replica to emulate the action of the buffer circuit shown in the transmitter schematic. This is intended to reduce sparking across the interrupter brushes where induced high-frequency pulses can produce enough back-EMF to result in this phenomenon.

Test results reveal that the spark-generated signal from the Wilson transmitter replica can be heard from the bottom of the broadcast band all the way up to 3.5 megahertz and above—which rather diminishes the notion of 'sharpness' in tuning. It is little wonder that when truly sharp tuning associated with valves and continuous waves became possible, spark signals were very quickly banished from the radio frequency spectrum for military communications purposes.

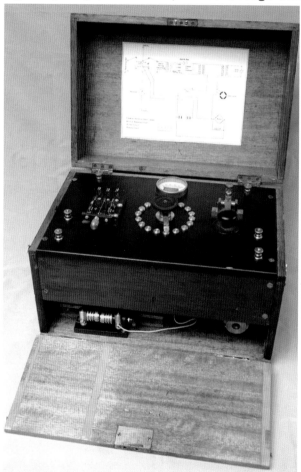

Replica of the Wilson transmitter showing spark gap (PRJ)

Part 2: 1920–1949

5 The Interwar Years

The war that had ground to its bitter end in November of 1918 may have been thought of as the 'war to end all wars', but the terrible reality was that subsequent events would lead inexorably to a continuation of the conflict within a historically short period. With the benefit of hindsight, it is apparent that fundamentally divergent German and Allied views of what had transpired in the last months of the First World War were major contributors to what was to follow.

While the German Army and the German population at large believed that a 'truce' or 'agreed cessation of hostilities' had occurred in November 1918, the Allies, and particularly the United States, considered that total victory had been achieved—a victory that allowed them to demand reparations and a revision of national boundaries that Germany's citizens would later find intensely objectionable. The Treaty of Versailles contained a range of punitive clauses which became the basis of national discontent and agitation that within a comparatively short time found expression in the new political force of National Socialism. Adolf Hitler, an Austrian soldier who had served in the First World War as a corporal, became the leader of the Nazi Party.

The Treaty of Versailles was not signed until June 1919 and laid responsibility for starting the war squarely at the feet of the German nation in a so-called 'guilt clause' that fuelled the tide of resentment that exploded into further world conflict in 1939.

Other elements of the Treaty of Versailles which the German nation found unacceptable included the annexation of Alsace-Lorraine by France, and the reassignment to France and Great Britain of all German colonial possessions. The assignment of Silesia, part of Germany, to the reinstated Polish nation and the imposition of enormous financial reparations (which were not finally paid off until the 1980s) added to national discontent during the 1920s.

A contributing factor to the cessation of hostilities in the First World War was the Russian Revolution in 1917. This led to the signing of the Treaty of Brest-Litovsk between Russia and the Central Powers (Germany, the Austro-Hungarian Empire, the Ottoman Empire and the Kingdom of Bulgaria), which recognised Russia's withdrawal from the Eastern Front and the end of its part in the war. The Treaty also required Russia to surrender significant areas along its borders that allowed the re-creation of states that had existed in historical times. These included Finland, Estonia, Lithuania and Latvia in the north-west, and Belarus and Ukraine in the south-east.

These events, and other steps intended to establish a binding peace and a stable relationship of national interests, instead generated a turbulent and unstable period of twenty years. These two decades were also marked by the development of highly significant telecommunications technologies, which had their genesis in the demands for improved message handling in the trenches of the Somme. Before very long, however, in Europe and the Far East these advances would be employed to reflect heightened political and national animosities and ultimately be coupled to the opening of fresh hostilities.

One of the more significant aspects of post-World War One thinking in the United States, Great Britain and France was a general revulsion for military matters and a desire for peace at almost any cost. This attitude was particularly marked in the United States, where strenuous efforts were made to ensure there would be no future involvement in wars that might occur in Europe, or in any conflict that did not directly affect the interests of America. This led to the policy of isolationism that only came to an end in 1941, when the United States was attacked by Japan at Pearl Harbor. In Great Britain and France this same attitude of 'peace whatever the cost' led to a policy of appeasement that continued almost up to the outbreak of the new war in 1939.

Communications Technology

In terms of technological development in communications, the most obvious manifestation of the concern to maintain the peace was ten years or more during which little happened in the military sphere and funding was severely curtailed. This was particularly the situation in Australia, where military wireless apparatus frequently constituted obsolete hand-me-downs from Great Britain or Australian-built versions of the same technology. Only after 1935 was it appreciated that in any future conflict Australia was likely to need to provide its own military communication requirements.

Public Broadcasting

While the emergence of radio broadcasting in the 1920s could not be seen as military in character, a fundamental component of broadcast radio transmission owed its existence to developments during the First World War which saw the thermionic valve converted from the temperamental device developed by De Forest in the form of the Audion into the hard vacuum triode used in trench radio systems. During the late 1930s, developments in the broadcasting revolution would be fed back into the design and development of military apparatus. In the United States this included the work of Edwin H Armstrong on his method of transmission and reception based on frequency modulation.

Airzone Cub (PRJ)

From the early 1920s, when radio broadcasting commenced in the United States, Great Britain, Europe and Australia, the new technology thrived as it responded to the demands of an enthusiastic public. More importantly, this growth demanded the development of increasingly sophisticated methods of reception as the radio frequency spectrum became ever more crowded with broadcast stations.

In the United States in particular, the growth of broadcast stations was encouraged by commercial exploitation of the medium. The development of business advertising utilising this new method of access to the public enabled intense competition to flourish. In this booming environment, the need to separate closely spaced radio stations demanded the use of the superheterodyne method of reception. The superhet became increasingly popular and led in turn to an increasing demand for the valves that this form of circuit required.

In its most economical circuit arrangement, the superhet required four

valves as compared with a radio that used a regenerative detector, which could achieve reasonable results with only two valves. However, apart from its ability to separate closely spaced radio stations, the superhet made a radio receiver a considerably easier device to operate than one that used the regenerative detector. Its popularity saw it become the ultimate radio receiver for general use up to the present day.

EKCO radio UK: the AD 65 designed by Wells-Coates, 1934 (INT)

Mechanical and Electronic Television

Although the notion that images could be conveyed to distant places as electrical signals had been mooted in the late 1880s, even before the work of Lodge, Braun and Marconi, it was not until the 1920s that the technology to support the theory became available. How electronic technology might be employed to create and transmit a television image had been proposed as early as 1908 by Alan Archibald Campbell-Swinton, the Scottish engineer who had promoted Marconi's early work in Great Britain. However, when John Logie Baird came to the idea of creating and sending images ('television') in 1925, he approached it in a fashion that was largely mechanical. In the Baird television system (the Televisor), a rotating disk was used to scan the image to be sent to a distant place rather than the electronic scanning system proposed by Campbell-Swinton.

Baird Televisor, 1929 (PRJO)

The Nipkow scanning disk had been proposed in the late 1880s by the German inventor Paul Gottlieb Nipkow as the basis of a television system to operate over land lines. The novelty of the Baird approach lay in the use of wireless communication to distribute the image rather than the creation of an image through electrical impulses, per se. It was not until the work of the independent American inventor Philo Farnsworth, and the Russian-American physicist Vladimir Zworykin, in the late 1920s, that true electronic scanning was achieved. Baird's mechanical and rotating perforated disk scanning methods were superseded by electronic scanning in Farnsworth's Image Dissector camera and the Iconoscope camera developed by Zworykin at Radio Corporation of America (RCA). These cameras were able to produce convincing, higher definition images that could be transmitted, and received at a distant location via a cathode-ray tube.

Marconi–EMI television, UK, 1935 (PRJ)

Whatever the value of television to the public's taste for news and entertainment, it was the application of electronic scanning in an entirely new navigational and defensive application that was to prove of critical importance to military activity at the end of the 1930s. Radio location and ranging was a concept that had its genesis as far back as 1904 in the ideas of the German inventor Christian Hülsmeyer, when he proposed to detect the presence of ships at sea through the reception of reflected radio energy pulses. However, it was not until 1934 that the military application of radio detection and ranging was considered by the British government as a possible response to the new high-speed aircraft that were being developed in Germany. RADAR, the acronym by which radio detection and ranging came to be known in the United States, and eventually worldwide, was developed in those few years before the start of the Second World War and proved an indispensable aid in responding to the massed attacks of the Luftwaffe during the Battle of Britain.

Radar was not a complete solution to anticipating air attacks and a device that had been developed during the latter part of the First World War was brought back into operation. This was the plotting table, a translucent map of the United Kingdom and the Continent lit from below, on which the position of enemy aircraft could be indicated. This permitted a comprehensive appreciation of the disposition of all combatants and enabled a very rapid and effective response by the Royal Air Force. Use of the plotting table also precluded the need to maintain flying patrols with associated pilot fatigue and excessive fuel usage and allowed the limited number of aircraft available to be used to best advantage.

Plotting table at RAF Command Centre (*ILN*)

Development of Radar

In the late 1930s, scientists in both Germany and Great Britain were working on systems of radio location, or radio direction finding (RDF) as it was known initially in England. Because the system developed in Great Britain was established on relatively low frequencies, around 20 megahertz, and the German system was an order of magnitude higher, in the vicinity of 200 megahertz, it was some time before the existence of the two systems became mutually apparent.

In 1939, Germany's last remaining rigid airship, the Graf Zeppelin II registered as *LZ 130*, was sent on a voyage around the shores of Great Britain to listen for radio location signals. Because of this difference in frequencies, the German radio operators failed to hear the experimental RDF radiation from the very visible and prominent RDF antenna towers. On the other hand, even though still at a somewhat experimental stage, the British RDF observers were able to plot the track of the Zeppelin around the eastern edge of England and up along the coast of Scotland.

Initially, what both the German and British RDF systems and the US radar lacked was a means of producing very short wavelength and high-frequency radiation that would make possible highly accurate distance and directional information. This problem was solved with the invention of the cavity magnetron by John Randall and Harry Boot at Birmingham University in 1940. With its capacity to produce centimetric wavelengths at powers of up to 10 kilowatts, in pulses, the device revolutionised the capacity of radar to provide navigational information. It became the heart of airborne radar as it was developed to assist night bombing raids over Germany as the war progressed, and it allowed the RAF and the US Air Force to detect German fighter aircraft during those raids.

In the period immediately prior to the outbreak of war, the British radar system was inherently less precise than the much higher frequency German system. Despite its limitations it proved adequate for the task. Detecting the location of incoming aerial intruders and providing an indication of the distance from the radar station was sufficiently accurate to allow the disposition of the aircraft to be displayed on the plotting table.

The information provided in the chain home RDF system developed

Chain home direction-finding antennas for transmit and receive (ILN)

at Bawdsey Manor, and used to plot the location of both intruders and friendly aircraft, enabled the control and husbanding of the resources of RAF aircraft required to repel the forces of the Luftwaffe. The use of the plotting table enabled a clear understanding of where the maximum effort should be applied and underlay the remarkable feat of defeating the German air force in the Battle of Britain. This British success resulted in Hitler's decision to postpone the invasion of Great Britain until German command of the skies could be ensured. This never happened.

6 Interwar Military Communications

Work on portable radio receivers and transmitters continued in the British Army after 1918, but progressively less and less financial resources were made available for weapons development. Thus two key weapons of the First World War, wireless and the tank, received progressively diminishing support until the mid-1930s, when German rearmament forced a complete change of attitude in Britain and France and a heightened response to the Nazi threat.

In this period Australia's dependence on the 'mother country' for its armaments and military support, particularly wireless, came under serious challenge. Wireless apparatus used by the Australian Army in the 1920s was at least two generations behind the technology available to the Royal Corps of Signals in Great Britain.

Portable Field Radio

The British Army radios used by the Royal Australian Corps of Signals in the post-First World War period were large and heavy and operated on the relatively long wavelengths in use during the war years. The primitive valve transmitters of this British equipment were powered directly by portable petrol generators or heavy lead acid batteries, resulting in a cumbersome and inefficient system with a relatively restricted communication range.

This gives the impression that Australia was seen as a distant and unimportant place where British Army surplus could be disposed of. It was Australian recognition of this perception that was to drive the political attitude that technological self-sufficiency was essential and saw many radio and valve industrial developments in the 1930s. Amalgamated Wireless (Australasia), or AWA, the largest Australian manufacturer of radio and electronics components at that time, created a new division,

Amalgamated Wireless Valve (AWV), which was accommodated in an enormous establishment in the Sydney suburb of Ashfield.

Tragically, the industrial might of AWA, as embodied in the AWV factory, has long since been lost, along with the technological capability of most radio and electronic manufacturing, and reliance on Asian manufacturing has become the norm.

The first generation of post-war British-sourced communications equipment was delivered as a wireless transmitter and receiver pair, the Type A Mark 2, referred to as the 'Ack' set. Like the much earlier Marconi pack set, this involved a wireless station that was very heavy and required a considerable team to transport it from place to place and to operate it. Here six signallers of such a group are shown with the set.

Type A Mark 2 'Ack' set—receiver (left) and transmitter (right) (RAS)

The size of the signals group required to transport and operate such early valve apparatus as the Type A Mark 2 is even more apparent in the illustration of its replacement, the Type C Mark 2. This set was generally referred to as the 'Cork' set. 'Ack' and 'Cork' were letters from the phonetic alphabet in use by the Australian Army at that time.

The Royal Australian Signals Corps received the Type C Mark 2 wireless in the late 1920s, and it remained in service until replaced by the Wireless Set (WS) No. 101 in 1939. Although made obsolete by the new set, which operated in the high frequency band, the Cork set remained in use by the Citizen Military Forces for a considerable period.

Type C Mark 2 Cork set: receiver, transmitter and high tension generators (RAS)

During 1929 and 1930, work commenced in Great Britain on a new transmitter and receiver pair based on the earlier experimental wireless set Type A Mark 3, and it was introduced as WS No. 1 in 1933. However, the new radio had a limited range of frequencies, 4.2 to 6.66 megahertz, and was also quite low powered when using continuous wave (CW) for the transmission of Morse code at 0.5 watts. This set could also produce radio telephony (RT), but in this mode the output was significantly reduced. With these limitations fully appreciated, work was commenced on a successor, Wireless Set (WS) No. 11, which became available in 1938.

In Australia, samples of the British WS No. 1 became the basis of a locally produced radio, superficially a clone of the British set. Known as the WS No. 101, it went into service at about the same time that the WS No. 11 became available in Great Britain. In another Australian modification the WS No. 101 was adapted to use an 807 valve to produce a higher power output—in this configuration it was known as the FS6 wireless set.

Australian signallers using British WS No. 1 (RAS)

The WS No. 101 consisted of a receiver in a compartment on the right and the transmitter in a separate compartment on the left in the one metal box, together with a power supply in a separate metal box. The valves for both parts of the radio set were contained in an elongated compartment along the bottom panel with a hinged lid to allow access and replacement. The Morse key in this same area was set on a sliding base that ran on rails.

This radio was supplied from a separate power supply that incorporated a vibrator to allow the creation of high voltage direct current to operate the valves. This power supply was operated from a 6 volt battery pack, carried either by hand or as vehicle equipment.

The FS6 was virtually identical to the WS No. 101 externally, with one major difference. The 807 valve was located immediately adjoining the port for the Morse key, with a small metal dome provided to allow for its extra length over the pair of 1K5 output valves used in parallel in the WS No. 101.

Considerable numbers of the FS6 were sold to the Indian Army, and it was also used by the British Army during the campaign against the Japanese in Burma. In northern Australia, the FS6 was used by Norforce when, early in 1942, the threat of Japanese invasion appeared to be imminent after the Pearl Harbor attack.

Tank Warfare and Wireless: *Blitzkrieg*

While it had proved a winning weapon for the Allies in 1917 and 1918, after the Armistice the military tank tended to languish. Somewhat ironically, it was the defeated German Army that saw the value of the tank and during the 1920s incorporated its development into its assumption that the future operations of a revived German Army would need to be highly mobile, with the tank as its principal component.

Among the controls on German military activity embodied in the Treaty of Versailles had been a severe limitation on battleships, the prohibition of the manufacture of tanks, and a numerical limitation on the size of the German Army at 100 000 men.

Upon his accession as German Chancellor in 1933, Hitler embarked on a process of rearmament and, in particular, sought ways to avoid the prohibition on tank construction and operations. Even before the hulls of a new generation of German tanks could be built, ways were found to train prospective soldiers in the art of tank manoeuvres. Soon, bizarre fleets of *Panzeratrappen*, wooden and cardboard replicas of tanks mounted on the chassis of motor vehicles, were secretly trundling around the open plains of northern Germany. By this means the simulation of the tactics of the tank attack was made possible, and so too was perfected the tactic of *Blitzkrieg* (Lightning War) in which the tank was the dominant weapon of attack, supported by following infantry.

An enthusiastic exponent of this new method of assault was the relatively young General Heinz Guderian, who had been a Signals and General Staff officer during World War One. His military communications background was instrumental in creating his interest in mobile, tank-based warfare that led to the creation of the Panzer Division. Wireless was the only effective means of directing and controlling this new mobile element of the German Army.

Wireless would become the critical element to the success of a rapidly advancing attacking force, and its integration into the Panzer forces was fundamental to the planning of the resurgent Wehrmacht when it launched the assault on Poland in 1939. Additionally, transferring messages to rapidly moving forces required secrecy and a new method of coding was required. A mechanical enciphering device known as the Enigma was the answer. This machine became not only the method employed by the German Army, Navy and Air Force to transmit and receive secret information, but through the decryption of its codes would also become the window through which Great Britain was able to view what its opponent was doing.

Mobile Enigma and General Heinz Wilhelm Guderian, leading proponent of tank warfare and mechanisation in the German armed forces (INT)

Signals Intelligence and Enigma

Perhaps anticipating that it was high on the list of nations that would bear the brunt of a Nazi invasion, during the inter-war years Poland had been at the forefront in developing techniques to respond to German coded messages, particularly the machine-generated code that had begun to appear in the late 1920s and early 1930s. It was assumed in Germany that this form of code was impossible to break into and thus completely secure. It was generated by an Enigma, a machine invented at the end of World War One by the German engineer Arthur Scherbius that employed multiple coding wheels. The Enigma certainly produced a coded output of massive complexity, but the Polish Cipher Bureau, deciding that the code might not be completely impregnable, employed a number of very bright young mathematicians to investigate the possibility of deciphering the messages.

Four-rotor naval Enigma (PRJ)

Using pure mathematics and statistics, the Enigma code was broken in the early 1930s. In part the solution was based on a mechanical analogue of the Enigma machine, with which the young mathematicians were able to analyse the most probable configurations of code wheels in a relatively short space of time. (This device was called the Bombe, based on its resemblance to a portable ice-cream making machine in use in Poland at the time.)

In this early period, the Enigmas were equipped with three wheels, the order of which could be changed when a machine was initially set up for operation. When the number of code wheels was increased to five in 1938, the number of possible arrangements increased enormously and the Polish Cipher Bureau was no longer able to decrypt German messages. Judging that hostilities were imminent, Poland invited France and Great Britain to take over the material that had been accumulated. This became the basis of the successful penetration of German secret messages, referred to in Great Britain as 'Ultra'.

The story of the British attack on the Enigma code after that generous action by the Polish Cipher Bureau has generated a significant body of literature since it was exposed by former RAF officer Frederick W Winterbotham in 1975 in his controversial book, *The Ultra Secret*. However, the first mention of work on the Enigma code appeared in the book by US journalist David Kahn, *The Codebreakers*, as early as 1966.

What is particularly relevant to the story of military radio is that not only did the German Army depend for its success on the supposedly indecipherable Enigma code but so also did the Luftwaffe and the German Navy. It was this widespread usage that allowed the Allies to access messages via a code-breaking group set up at Bletchley Park, the site of the Government Code and Cypher School.

Bletchley Park: main building at Station X (PRJ)

To gather this material, the British authorities established a corps of wireless message interceptors known as the Y-Service, an organisation that remained shrouded in mystery long after the Ultra secret had been exposed. The Y-service was composed of former radio amateurs and civilians from a telecommunications background who had been recruited to undertake this highly secret activity. The 30-year silence that followed the end of the war was in many cases never broken; in some instances, married couples who both worked at Bletchley Park (Station X) were quite ignorant of each other's presence at the time. Perhaps this is not so surprising when one considers that the population of the Bletchley Park estate grew to some 10 000 persons by 1945, the size of a small town.

The counterpoint to the photograph of Guderian supervising his signallers in a tracked signal van is the photograph opposite of the receiver used in large numbers by Y-Service operators at the receiving stations in

Great Britain; these were provided by the United States as part of the Lend Lease support prior to US entry into the war in 1942. This photograph shows a typical listening post for a Y-Service interceptor, recreated at Bletchley Park. The principal item of equipment is the receiver manufactured by the US National Radio Company, commonly known as the National HRO. (HRO was said to stand for 'helluva rush order'.)

Rather than different reception ranges being obtained via a system of switching, this radio employed plug-in coil boxes to cover a particular range, with the result that a highly stable receiver was available with negligible frequency drift. Where the need to return to a particular frequency was of paramount importance to the success of interception, this facility made the HRO an indispensable tool.

Radio amateur G4GAN (SK) operates an HRO at Bletchley Park (PRJ)

7 World War Two

When the stock of a present-day bookshop is examined and the extent of the section dealing with the Second World War is fully appreciated, one might be forgiven for assuming that everything that could possibly be said about this conflict has already been said. This opinion tends to overlook the endless fascination of this period in which so many people from all parts of the world were drawn into a brutal five-year conflict.

Very few of those who participated directly in World War Two remain to recount their stories and it is left to the generations that follow to take up the tale. Compared with earlier historical undertakings, the task is made easier by the existence of a huge body of written material, coupled with an enormous volume of photographic and cinematic recorded evidence, the latter very often available in remarkably clear, well-preserved high-definition images.

One of the first things that can be said about World War Two is that telecommunications from the outset was a vital element. Whether on the battlefield, in the air or at sea, wireless (by then termed 'radio') provided essential communications and later a means of conveying secret messages.

Radio broadcasting was the means by which the declaration of war was announced by Neville Chamberlain in Great Britain in 1939. And in 1945, the Japanese Emperor announced the end of the war in the Far East by radio. During the course of the war, Lord Haw Haw (William Joyce) and Tokyo Rose (in fact, a number of different English-speaking Japanese women broadcasters) were the spearheads of Axis broadcast propaganda. This could be contrasted with the BBC, which strove to present truthful reporting, even when it was painful for Allied listeners.

Anticipating German Aggression

During the later 1930s, for a few clear-eyed commentators Germany's ambition as expressed by Adolf Hitler, leader of the National Socialist Party, became increasingly clear—European domination. In Great Britain, a politician who had never completely recovered from the disaster of Gallipoli employed the time made available by withdrawal from public life in writing books. During this period, Winston Churchill also worked to galvanise public concern about the resurgent German nation and its gathering military might. How this former First Lord of the Admiralty came back from obscurity in his sixties to lead the country when all hope seemed lost is very well known. With the German Army poised on the coast of France to cross the English Channel and invade Great Britain, he inspired the nation to resist by all means.

In an era when the whole world is aware of military activity, however distant, that Churchill's principal means of inspiring the British nation was via the broadcasting services of the British Broadcasting Corporation is something that the current generation of Twitterers and smartphone chatterers may find hard to imagine. What is less well known is that as a child Churchill was cursed with a stammer that he was eventually able to overcome. This success allowed him to present some of the most inspiring and memorable speeches ever uttered by a politician, speeches that were recorded and can still be heard long after the event. A Google query will take an interested reader to the best known of these speeches.

Assault on Poland

During the late 1930s, Hitler's plans for the domination of Europe saw a series of events that had the capacity to reignite the conflict of 1914–18. However, when his ambition led to the annexation of Czechoslovakia in 1938, the politics of appeasement drove the British Prime Minister, Neville Chamberlain, to a hasty attempt to head off conflict and achieve 'peace in our time', an action that provided all the encouragement the Fuhrer could have required to invade Poland. Perhaps too, knowledge of the infamous resolution of the Oxford University Union Debating Society in 1933 helped to encourage his aggressive stance in dealing with Great Britain and France. The resolution 'That this House will in no circumstances fight for its King and Country' may well have helped to convince Hitler he was faced by a degenerate and timid nation that could be bullied into submission by high-flown and exaggerated rhetoric. A bad error.

On 1 September 1939 German Panzers rumbled across the Polish border and Great Britain, to honour an undertaking to support this vulnerable nation if were attacked, declared war on Germany at 11.15 am on 3 September. It has been speculated that this was an outcome that Hitler had expected to avoid. However, the fateful step had been taken, and once again a state of war existed in Europe. Australia once again rushed to the aid of the mother country, perhaps heedless of the perils developing closer to home in an area that lay under the acquisitive gaze of a militaristic and nationalistic new player on the world stage, Japan.

As the voice of Chamberlain filled living rooms across Great Britain and in France, many old soldiers must have been filled with dread for the future of their children as they remembered the horrors of that earlier conflict. Chamberlain's somewhat dispiriting opening words were:

> I am speaking to you from the Cabinet Room at 10, Downing Street.
>
> This morning the British Ambassador in Berlin handed the German Government a final note stating that unless we heard from them by 11.00 a.m. that they were prepared at once to withdraw their troops from Poland, a state of war would exist between us.
>
> I have to tell you that no such undertaking has been received, and that consequently this country is at war with Germany.

And he concluded:

> Now may God bless you all. May He defend the right. It is the evil things that we shall be fighting against—brute force, bad faith, injustice, oppression and persecution—and against them I am certain that the right will prevail.

With the German attack on Poland progressing on its brutal way, an eerie calm descended over Europe for some months and the phrase 'phony war' was in the air. However, the German U-boats were at sea and soon Britain's HMS *Royal Oak* was sunk. The news of this disaster was gleefully announced over the airwaves by Lord Haw Haw. It would be a number of years before this malignant radio personality would be brought to justice—the Irish-American fascist politician William Joyce was eventually tried for treason in 1946 and hanged at Wandsworth Prison, aged 39 years.

During the period of phony war, the German Army was systematically destroying the Polish military forces, which included large numbers of mounted cavalry and obsolete biplanes. The ancient string and wire aircraft of the Polish Air Force were sent up to confront an enemy equipped with high-speed monoplanes and employing tactics honed in the skies over Spain during the civil war of 1936–39 that elevated General Franco to a dictatorship that would last until 1975. Germany had also been able to test the newly constructed tanks of the Panzer forces in Spain. However, when Hitler pressed Franco to join the Nazi invasion of Europe he was refused, and Spain remained neutral during World War Two.

Miracle at Dunkirk

By 10 May 1940, as the Panzer army raced across France in a pincer movement that began from the Ardennes in the south and across neutral Belgium in the north, the only question that could be asked was: 'Could this force be stopped?' And the short answer was 'Yes'—but only by the English Channel.

The rapidity and invincibility of the pincer movement forced the British Expeditionary Force and the French Army back into a restricted pocket at Dunkirk. From this confined battleground, the majority of the British forces and some of the French Army were saved by the extraordinary efforts of a fleet of small ships that sailed across the Channel to evacuate them. Virtually all arms, artillery and vehicles had to be abandoned, but the men survived to fight another day. Hitler's 'final attack' was arrested.

Why Germany held back at this point remains one of the unexplained mysteries of World War Two but perhaps it was because, at heart, Hitler had hoped to avoid an ultimate confrontation with Great Britain. Perhaps, too, he perceived that such a step would inevitably draw the United States into the conflict, just as in the First World War. Moreover, invasion required the transfer of German troops across a stretch of water commanded by the ships of the Royal Navy. Faced with this prospect, Hitler relied on the new German Air Force, the Luftwaffe, as his primary weapon of invasion. Thus it was in the skies over southern England that the Battle of Britain commenced in early July 1940, with a relatively tiny band of airmen battling to arrest the German onslaught.

Australia Joins the Fray

Geographic reality dictated that the Australian military contingent, embodied in the Second Australian Imperial Forces, would not be available for direct action in the new European conflict for some months. However, early Australian participants were around thirty pilots, who joined the Royal Air Force in time to participate in the Battle of Britain. By 1945, the Australian military personnel in Europe, in the main serving in the Royal Air Force or attached through the Royal Australian Air Force grouping, were quite substantial in number. By the end of the war, in Bomber Command alone, the number of Australian fatalities over Europe was nearly 3500—although this number represented only 7 per cent of the overall Bomber Command death toll.

Australian troops were again deployed to the Middle East, to assist in opposing the Italian incursion into Egypt from neighbouring Libya. Very much as had occurred in World War One, Australia went to war desperately under-prepared following twenty years of economic stringency. This was particularly true of the signals and communications facilities, which had only recently been seen as requiring a significant investment in skill and effective portable radio communications equipment.

The Australian forces that arrived in Egypt in December 1940 had only their training equipment available, and it was not until the middle of 1941, during the siege of Tobruk, that at last 17 of the required 76 radio sets, WS No. 108, were received. When further WS No. 108 sets and the newer WS No. 109 sets were received in August 1941, it was apparent that some deliberate sabotage had occurred, either at the time of manufacture or prior to despatch from Australia by sea. This required repair and further delay before the new equipment could be delivered.

With the deployment in North Africa of German troops under General Rommel after February 1941, a campaign of shifting fortunes developed along the Mediterranean coast. A major component of the Australian forces was moved to Greece to join the British Army in what turned out to be a hopeless campaign, in which the Allies were both outnumbered and

Wireless set WS No. 108 Mark 1 (PRJ)

Wireless set WS No. 109 Mark 2 (PRJ)

outmanoeuvred. In a relatively short period, the German Army successfully evicted the Allied forces from Greece, from which they retreated to Egypt or became involved in the defence of Crete. This latter campaign also proved to be a disaster in which many Australian and New Zealand lives were lost; the survivors were captured and taken north by rail in cattle-cars to imprisonment in Germany for the duration of the war. The German paratroops did not achieve victory unscathed, however, their losses being so extreme that Hitler would never again employ this form of assault.

Following initial major Allied setbacks in North Africa, in which the capture of Cairo looked likely, General Montgomery was appointed to respond to the successes of the German Eighth Army under General Rommel, a highly competent and well-trained opponent. While Montgomery had access to Ultra intelligence, which provided an understanding of German strategic intentions, it transpired that Rommel in turn had been well advised about the future intentions of the Allies. The spying activities of Elyesa Bazna, the trusted Albanian valet of the British Ambassador in Ankara, Sir Hughe Knatchbull-Hugesson, provided Rommel with information which underpinned a number of his most critical decisions in the North African campaign.

Known by the codename 'Cicero', this enterprising spy obtained a fund of high-grade intelligence as a result of the Ambassador's lax security. While it has been suggested that the British Double X Committee may have had a hand in providing misleading information, it appears that Cicero's information had significant impact on the success of Rommel's North African campaign. When the spy's identity was finally revealed through Enigma decodes, Bazna hastily left the Ambassador's employ, to later discover that his payments from the German High Command had been made in forged British Pound Sterling notes—a considerable irony.

The Asia and South-West Pacific War

With the unannounced attack on Pearl Harbor by the Japanese Navy and Air Force on 7 December 1941, Australia found itself in a war zone for which it was desperately unprepared—and very short of troops. With the majority of the infantry and associated gunnery and tank forces involved in North Africa and the Middle East, a state of public panic became perceptible in Australia. This quickly escalated with the fall of Singapore following the earlier sinking of HMS *Prince of Wales* and HMS *Repulse*.

During the previous year there had been a change of both leadership and government in Australia. Robert Menzies, who had spent rather too much time in Great Britain and away from the political pressures building up in Australia, was replaced by the leader of the Country Party, Arthur Fadden, in August 1941. However, the United Australia Party that Fadden led, involving Liberals and the Country Party, was defeated in October of that year and a new Labor government was elected, led by John Curtin. Curtin was far less Anglophile in orientation than Menzies, and when the threat of Japanese invasion appeared imminent he was far less inclined to respond positively to British demands for continued military support on the other side of the world. Following the tremendous Allied success in the Second Battle of El Alamein in October 1942, a substantial element of the Australian forces in that part of the world was withdrawn to take part in the war in the South-West Pacific area and support the small contingent of poorly trained and badly equipped troops remaining in mainland Australia.

The debacle of Singapore, which could be significantly ascribed to the poor leadership and inappropriate tactics of the British leader, General Percival, led to an ignominious surrender to a brutal enemy on 15 February 1942, and the subsequent imprisonment of nearly 15 000 Australian troops. The Australian military leader, General Gordon Bennet, escaped to Australia with other senior officers, leaving his men to the mercies of the Japanese.

At this time it became quite apparent that Australia's national interests lay in the Pacific area, and most particularly in concert with those of the United States. With Japan as the common enemy, bonds were forged with the United States which have persisted to the current time. Nonetheless, at an economic and business level ties to the United Kingdom have remained, although considerably lessened over the intervening years.

Indicative of the general level of panic that engulfed Australia following the Japanese attack on Pearl Harbor was the inaccurate allegation by Labor minister Eddie Ward that the previous government (under Menzies and Fadden) had been prepared to abandon the northern part of the mainland—above the so-called 'Brisbane Line'—in the face of a Japanese invasion. Indeed, with the partially successful government cover-up of the attack on the port of Darwin on 19 February 1942, only a few days after the fall of Singapore, it was a quite reasonable assumption that Australia was the next target for a Japanese invasion and therefore abandonment of the northern part of the continent might be expected to occur. That such an invasion never occurred is now directly associated with the Battle of the Coral Sea in May 1942, in which Japanese naval forces were conclusively defeated by the US Navy. In addition, in the months following the fall of Singapore, the Australian Army was massively expanded so that by June 1942 it included ten infantry divisions, three armoured divisions and myriad supporting units. The RAAF and the RAN had also been expanded, and new citizen forces had been developed, among them the Volunteer Defence Corps and the Volunteer Air Observer Corps.

With the major victory at Midway Island in June 1942, the threat of imminent invasion was averted, although it took almost a year for this to be appreciated by the Australian government. In the meanwhile, Japan had substituted a strategy of encirclement and isolation of Australia and its military forces. The Allied response to this strategy was to wage a series of campaigns in Timor, Indonesia, Borneo, Papua New Guinea and other parts of the South-West Pacific. US General Douglas MacArthur was given control of all of the available Australian and US forces in the South-West Pacific theatre of war, which had the effect of relegating General Blamey, the Australian supreme military commander, to a subsidiary role.

The Northern Campaign

Reference to a map of Australasia reveals the proximity of continental Australia to its tropical neighbours to the north, among them the island of New Guinea and the Malay Archipelago, a huge chain of islands which encompasses Indonesia, the Philippines, Singapore, Brunei, Sabah, Sarawak and East Timor. In a change of strategy following the defeat in the Battle of the Coral Sea, Japan decided to isolate Australia from American forces as far as possible, and an attack was launched on New Guinea in July 1942. A direct attack on Port Moresby in the south was untenable, so an assault was mounted at Buna in the north. There followed an epic struggle as the Japanese advance over the Owen Stanley Range via the Kokoda Track was staunchly resisted by a numerically inferior infantry group known as Maroubra Force, which consisted initially of the Papuan Infantry Battalion, and the Australian 39th Battalion, largely raised from members of the CMF.

Slogging over the extreme terrain of the Owen Stanley Range, by the middle of September Maroubra Force had been gradually pressed back to within visual distance of Port Moresby where, after a short delay, it was expanded by the addition of new regular army troops of the 53rd Battalion. At this juncture Maroubra Force was relieved by the 16th and 25th Brigades, and the subsequent counter-offensive drove the Japanese back to their beachheads at Buna, Gona and Sanananda on the north coast. US and Australian forces were not able to overcome Japanese resistance until January 1943, following attacks that involved very heavy casualties associated with headquarters pressure to achieve a 'quick' result.

Towards Armageddon: Atomic Conclusion to the War

With the progressive advances of the US Navy and Army northwards, ever closer to Japan, Australia's involvement in the conflict lessened to the extent that military expansion planned for in 1943 did not take place.

Unbeknown to almost the whole of the world at war, the pathway to the ultimate weapon was being forged in the High Sierras in the southern United States, and the name Los Alamos would forever be associated with the development of the atomic bomb and later the hydrogen bomb. Here J Robert Oppenheimer and a team of dedicated scientists carried out the work required to create this most awesome of weapons—which the Allies had striven hard to ensure Nazi Germany would not produce. The destruction of the Norsk Hydro factory near Rjukan in Norway, and with it the heavy water that it was forced to supply to Germany (discussed in Chapter 9), was but one of the operations designed to prevent the development of a Nazi nuclear device.

In the aftermath of the nuclear explosions over Hiroshima and Nagasaki, Australia became part of the army of occupation, in which General MacArthur was installed as something tantamount to a new emperor, although Japan's Emperor Hirohito was not removed or charged with war crimes, as might well have occurred under a less enlightened occupying force.

The ending of the war in Europe, with the capture of Berlin and the suicide of Hitler, followed by the surrender of the Japanese Empire on the orders of its Emperor, did not see an untroubled peace develop, however. Russian communism rapidly replaced Nazi tyranny as the principal opponent of Allied Europe and the United States. Now, in the words of Winston Churchill, an Iron Curtain fell across Europe that for decades would trouble the world with the potential for a nuclear war. Indeed, even more than sixty years on, the threat of nuclear holocaust remains as a potentially lethal outcome of the Second World War.

8 Radio Communications in Australia's War Zones

Just as had occurred in the First World War, the outbreak of the Second World War generated an immediate and pressing need for lightweight, compact, portable communications apparatus. Despite its isolation, over the previous decade Australia had developed a considerable radio production capacity to service the burgeoning broadcasting industry and in short order was able to produce radio transmitters and receivers that initially met military requirements. But in the humid tropical climate of the northern region that was now the operational environment, apparatus that was designed for conventional temperate zone conditions developed major problems. The ingress of damp and the formation of mould on internal components required new construction methods that later would become the norm for military communications apparatus.

Military Portable Field Radio

In the late 1930s, Australian military requirements saw the production of two radio transmitters, Wireless Set (WS) No. 101 and WS No. 109, the former based on the equivalent British WS No. 1, the latter on the British WS No. 9. The outbreak of war led to the creation of a radio that replaced both the WS No. 101 and WS No. 109 and bore the same number as another British radio of that period, the WS No. 11.

As discussed in Chapter 9, this radio was originally designed by the British Signals Experimental Establishment (SEE) and had some useful and important design features but also suffered at least two major problems—it was very heavy, and it was not adequately proofed against the ingress of moisture. These problems were perpetuated in the Australian version when it was produced by AWA in 1941.

Ostensibly a portable man-pack radio, in reality the WS No. 11 was

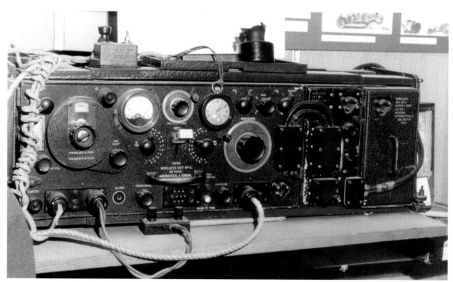

Wireless set No. 11 (Aust.) (PRJ)

Jeep-mounted wireless equipment in the tropical north, including a WS No. 133 (RAS)

only suitable for battlefield use when a vehicle or even a pack animal was available to carry it. The problem of moisture incursion was perhaps the more serious issue, because the high voltages required to drive valves did not mix happily with damp and internal condensation. In a tropical climate, the WS No. 11 was highly susceptible to fungal attack and frequently suffered component breakdowns, spending a considerable amount of time in the workshops being repaired or modified to improve performance. Despite these problems, it provided a useful intermediate step to its successor, the WS No. 122, which featured lightweight construction and was waterproofed to a standard that would allow short-term full water immersion.

British and US Influences on Australian War Radio

While involved in the campaign in North Africa, Australian signallers and infantry had become aware of WS No. 19, a radio recently developed in Great Britain for communications between tanks, armoured fighting vehicles and headquarters. The WS No. 19 included a compact VHF transceiver for tank-to-tank communications and an audio amplifier for

WS No. 19 with power supply, tuner and amplifier above (PRJ)

soldier-to-soldier conversations within the noisy internal tank environment.

Although this radio was not available for supply to Australia, its replication proved rather simpler than had been the situation with the WS No. 11 because the US-manufactured valves used in its construction were directly comparable with valves made in Australia by Amalgamated Wireless Valve. It was perhaps inevitable that the parent company, AWA, would be given the task of producing an equivalent radio for use in Australia and the near northern tropics. The result was the WS No. 19 Mark 2 (Aust.), which was supplied in 1943. Mindful of the difficulties experienced with the WS No. 11, AWA engineers ensured that the new radio was far better protected from humidity and moisture incursion. It was very successful—but also very heavy at 40 pounds (18 kilograms).

The need for a lighter and more compact field radio than the WS No. 19 led to the replication of another British radio, the WS No. 22, which had been used in the Battle of Arnhem (September 1944). Given the serious problems this rather underpowered radio had caused in that campaign, its adoption as a model was perhaps surprising. The Australian improved version, known as WS No. 22 (Aust.) appeared in 1945 near the end of the war and like the improved version of the No. 19 featured hermetic sealing around the edge of the front panel and a capacity to be immersed in water for a brief period, something that would have proved quite destructive in the earlier radios such as WS No. 101 and No. 11. The WS No. 22 (Aust.) was quickly followed in the same year by WS No. 122 (Aust.)

WS No. 22 as used at Arnhem (PRJ)

WS No. 122 (Aust.) initially was designed with a power output of 1.5 watts continuous wave; in its upgrade to the Mark 2 High Power version output was raised to between 6 and 7 watts, with a significant impact on maximum operating range using ground-wave propagation. Using Morse code and CW, the WS No. 122 Mark 2 High Power had a range of approximately 40 miles (over 60 km). Perhaps for this reason, it remained in service for at least ten years after World War Two.

WS No. 122 Mark 2 High Power (PRJ)

By the time WS No. 122 Mark 2 High Power was introduced, the impact of association with the US Army was starting to be felt and the need for compatibility with the radio apparatus used by this ally became obvious. Initially this led to the adoption of perhaps one of the most ubiquitous field radios of World War Two, the Handie-Talkie, with the designation BC-611. It was developed in 1940 by Galvin Manufacturing (now Motorola), and went into mass production in 1941.

The BC-611 was little bigger than a conventional telephone handset and was perhaps the simplest device to operate produced then or since. All that was required to activate the Handie-Talkie was to extend its telescopic aerial, which was coupled to a power switch. A pressel switch on the side of the case was depressed to speak, and as there was only a fixed frequency of operation no special skill was required of the operator. Perhaps the main drawback of this radio was its high frequency operation, between 3.5 and 5.995 megahertz. Because of the relatively short aerial, at this frequency the BCC-611's small power output was inevitably minimised by the inefficiency of the arrangement and the range that could be achieved was little more that 'line of sight' or between 2 and 5 miles (3.2 to 8 km)—but very useful nevertheless.

US design influences could be seen in the later man-pack field radio WS No. 128, which was produced during 1945 but not issued until the end of the war. This radio, while maintaining its place in the HF spectrum between 2 and 4.5 megahertz, in terms of its external configuration and battery compartment was closely related to the contemporary US Army man-pack field radio, the BC-1000. This latter radio operated in the

BC-611 Handie-Talkie (PRJ)

Australian WS No. 128 (PRJ)   US BC-1000 Walkie Talkie (PRJ)

VHF part of the spectrum between 40 and 48 megahertz and had a whip antenna which constituted a significant length relative to the frequency of operation; this was a highly efficient radio and achieved superior results from a minimal level of RF power output. More importantly, its use of the VHF range of frequencies and frequency modulation (FM) was a forerunner of a wholesale change in communications technology that took place in both the British and Australian armies. From now on, HF generally would be used for long-range, skywave-based communications, while tactical short-range communications would employ VHF and, later, UHF frequencies.

After the Second World War: Iron Curtain Impacts

The euphoria engendered by victory in the West and in the East was not to last long. As much of the apparatus and paraphernalia of war was disposed of at 'give away' prices, in the political world the new enemy, Communism, whether Russian or Chinese, would generate a series of localised military actions and new international tensions would see a fresh round of military communications developments. Though not of the scale of the recently ended World War Two, these military actions were significant and potentially only a step away from all-out nuclear conflict. The first was the outbreak of hostilities in Korea in June 1950 (discussed in Chapter 11).

In the meanwhile, as Australia wound down its military establishment, military communications apparatus was mostly the leftover equipment of the Second World War. Typical of this period and into the 1950s was the last British-inspired HF transceiver of the war, the WS No. 62, which bore a close resemblance to WS No. 122, a derivative of the No. 22 set produced in 1942. Reference to earlier illustrations will confirm this superficial resemblance when compared to the No. 62 set illustrated here

As in the WS No. 122, in the WS No. 62 tuning of both transmitter and receiver could be achieved with a single knob and depended on mixing the receiver local oscillator signal with a locally generated frequency equal to the intermediate frequency. Mixing of these two signals resulted in a transmitter signal at the same frequency as the received signal, obviating the need for netting. Crystal control of this transceiver was also available and again the frequency offset was applied, which in the No. 62 set was 455 kilohertz. As in the No. 122 set, two preset positions on the main tuning dial were available to enable rapid resetting of frequency while maintaining the original frequency and access to the military group.

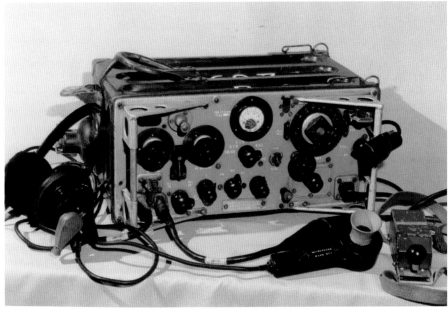

WS No. 62 manufactured by Pye Radio Ltd in 1945 (PRJ)

9 Clandestine Communications

In 1939, in the context of military and clandestine radio, what was developed and implemented in Australia was significantly circumscribed by the need to conform with the apparatus being used by the British Army and by Military Intelligence 6 (MI6), otherwise known as the Secret Service.

A major factor influencing the conduct of a war that was global in its complexion was the time it took to travel the great distance between Europe and the antipodes (and vice versa). While information could make the journey at the speed of light as radio waves beamed news of the war in Europe to Australasia and Southeast Asia, technological artefacts and people were faced with a long and perilous journey in the face of submarine warfare and the possibility of torpedo attack.

Because of these constraints, Australia established organisations to deal with secret warfare that mirrored the operations in Britain, while logistical issues and the fundamentally different environment in which radios were required to perform led to communication solutions that differed in a number of respects. The existence of advanced industrial communications enterprises in Australia allowed a ready response to local clandestine and military radio needs.

Clandestine Operations: MI6 and SOE

During the First World War, providing secret agents with a means of communications involving radio waves was in principle completely impossible while the method of generating a signal was the spark. Even with the advent of the hard vacuum valve, 'portable' communications apparatus remained either too heavy and bulky to be practical for clandestine operations or too low powered to cross useful distances.

The signal from the powerful German longwave station at Nauen, west of Berlin, was capable of reaching South America and South Africa, and the distribution of clandestine messages from this source was possible in principle, although return intelligence had to rely on concealment in letters with invisible ink or the use of open codes in telegraph messages. Fully encrypted messages to Germany via the telegraph service were of course not useful, as they inevitably marked themselves as suspect and likely to lead to the identification of an enemy agent.

What changed in the 20 years between the two wars was the progressive miniaturisation of valves and the capacity to reduce the weight and bulk of radio apparatus. Even in 1939, however, the British Secret Service was forced to send its agents into Europe with frankly ridiculous apparatus that must have signalled clandestine intent like a flaming beacon. The Mark 15 transmitter and receiver pair available just before the start of the Second World War were typical of this problem, being inordinately heavy and clumsy and contained in large wooden boxes. With the intervention of the new prime minister, Winston Churchill, and the creation of the Special Operations Executive (SOE), that situation soon changed for the better.

Whaddon Mark VII in leather carrying case (PRJ)

Type 3 Mark 2 in leather carrying case (PRJ)

A radio transmitter and receiver produced by the Radio Security Service as a replacement for the bulky Mark 15 equipment was the Whaddon Mark VII, later referred to as the Paraset when it was adopted by SOE for its first generation of aircraft-inserted agents. Although remarkably compact for that period, the Paraset had some most unfortunate characteristics that made field use extremely problematic. As discussed in the Part 2 project (page 175), the receiver circuitry was obsolete and employed the Armstrong regenerative detector circuit, invented before the First World War. A dangerous by-product of this was that in CW reception mode the Paraset receiver produced a low-level radio frequency signal that was capable of being detected by German interception teams.

Nevertheless, with output power of 5 watts or more, the Paraset allowed communication between France and Norway and the reception stations in Great Britain, assuming that a reasonable antenna could be erected. Away from built-up city areas, interception was less of a problem and the Paraset was used to a considerable extent in Norway for reporting weather conditions and the location of German merchant vessel convoys. Even more critically, the locations of the notorious battleships *Scharnhorst* and *Gneisenau* were reported as they sailed out into the North Atlantic to harass Allied shipping.

The relatively low-powered Paraset was followed by the generally superior Type 3 Mark 1 and the very popular Type 3 Mark 2. Both radios used superhet receivers and employed the 6L6 valve to produce a useful 20 watts of CW radio frequency energy. Because of these characteristics, in the post-war period the Type 3 Mark 2 became very popular with radio

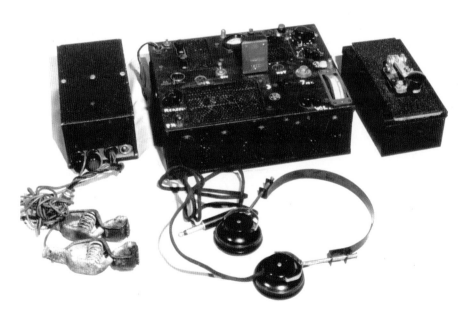

Type A Mark 3 clandestine transceiver, 1945 (PRJ)

amateurs looking for equipment to replace the radios surrendered at the start of the war. Towards the very end of the war, the pinnacle of SOE radio art was produced as the Type A Mark 3 (illustrated), which even today appears surprisingly compact, considering that it employs valves rather than transistors or other solid-state devices.

In the post-war period, many examples of this radio appeared in Australia, where they had been despatched as British government war surplus. Not surprisingly, this radio also found immediate favour with amateur operators to replace the radio stations surrendered in 1939. The Type A Mark 3 was also used by a number of government and quasi-government organisations such as the Forestry Commission and Bush Fire Brigades.

Atomic Weapons and National Socialism

In the late 1930s, scientists in Germany made the initial discoveries that would lead to the development of the atomic bomb. On 17 December 1938, Otto Hahn, a leading experimenter in nuclear physics, with his research associate Fritz Strassmann, succeeded in splitting the atom of uranium. This event was advised to their research colleague Lise Meitner, who had recently escaped the anti-Jewish environment of Germany to live in Switzerland. Meitner analysed their work and correctly interpreted the experiment as having involved 'nuclear fission', a term that persisted during the nuclear age that was to follow.

On 10 February 1939, in a second publication relating to nuclear fission, Hahn and Strassmann predicted the liberation of additional neutrons during the fission process. This, the most critical step towards the creation of self-sustaining nuclear fission that drives the atomic bomb, was subsequently demonstrated by the French physicist Jean Frédéric Joliot in March 1939. The additional fission products produce a cascading and expanding surge of fission products which, if not contained or moderated, lead instantly to a nuclear explosion. In the early days, the most important 'moderator' was found to be heavy water, in which some atoms of hydrogen are replaced with atoms of deuterium. Without going into a complex explanation of the chemical nature of heavy water, suffice it to note that an iceblock of heavy water sinks whereas normal frozen water floats.

Heavy water came to be the preferred 'moderator' for the creation of nuclear fissile material in Nazi Germany, and became of critical interest to the Allies in the process of developing their own nuclear fission technology as part of the Manhattan Project. Fortunately for the scientists involved, when the Nazis invaded France in 1940 Joliot smuggled his working documents and materials to England.

In the letter reproduced here, from Albert Einstein to the President of the United States, Franklin Delano Roosevelt, the Allies were alerted to the potential danger of nuclear experimentation in August 1939. Later, the implications of experimentation by the Nazis and the possibility that Hitler was intent on creating a German atomic bomb also became apparent.

Albert Einstein
Old Grove Rd.
Nassau Point
Peconic, Long Island
August 2nd 1939

F. D. Roosevelt,
President of the United States,
White House
Washington, D.C.

Sir:

Some recent work by E. Fermi and L. Szilard, which has been communicated to me in manuscript, leads me to expect that the element uranium may be turned into a new and important source of energy in the immediate future. Certain aspects of the situation which has arisen seem to call for watchfulness and, if necessary, quick action on the part of the Administration. I believe therefore that it is my duty to bring to your attention the following facts and recommendations: In the course of the last four months it has been made probable through the work of Joliot in France as well as Fermi and Szilard in America—that it may become possible to set up a nuclear chain reaction in a large mass of uranium, by which vast amounts of power and large quantities of new radium-like elements would be generated. Now it appears almost certain that this could be achieved in the immediate future.

This new phenomenon would also lead to the construction of bombs, and it is conceivable—though much less certain—that extremely powerful bombs of a new type may thus be constructed. A single bomb of this type, carried by boat and exploded in a port, might very well destroy the whole port together with some of the surrounding territory. However, such bombs might very well prove to be too heavy for transportation by air.

The United States has only very poor ores of uranium in moderate quantities. There is some good ore in Canada and the former Czechoslovakia while the most important source of uranium is Belgian Congo. In view of the situation you may think it desirable to have more permanent contact maintained between the Administration and the group of physicists working on chain reactions in America. One possible way of achieving this might be for you to entrust with this task a person who has your confidence and who could perhaps serve in an unofficial capacity. His task might comprise the following:

  a) to approach Government Departments, keep them informed of the further development, and put forward recommendations for Government action, giving particular attention to the problem of securing a supply of uranium ore for the United States;

  b) to speed up the experimental work, which is at present being carried on within the limits of the budgets of University laboratories, by providing funds, if such funds be required, through his contacts with private persons who are willing to make contributions for this cause, and perhaps also by obtaining the co-operation of industrial laboratories which have the necessary equipment.

I understand that Germany has actually stopped the sale of uranium from the Czechoslovakian mines which she has taken over. That she should have taken such early action might perhaps be understood on the ground that the son of the German Under-Secretary of State, von Weizsäcker, is attached to the Kaiser-Wilhelm-Institut in Berlin where some of the American work on uranium is now being repeated.

Yours very truly,

A. Einstein

(Albert Einstein)

Given their knowledge of Germany's potential development of an atomic device and the significance of heavy water in the production of fissile material, the source of this rare material became of major concern to the Allies. The principal source of supply was the isolated Vemork hydro-electric power plant developed by Norsk-Hydro in 1911, near the small town of Rjukan (pronounced Roo-kun) in occupied Norway. This power station generated electricity for industrial processes in nearby Rjukan but in the 1930s it also provided cheap electric power for the creation of hydrogen gas by the electrolysis of water. Hydrogen was an essential component of the fertiliser which was the intended end-product of this process. Electrolysis also resulted in the production of heavy water (a form of water that contains a larger than normal amount of the hydrogen isotope deuterium) which is present in ordinary water in minute quantities (1 part in 6000).

The Vemork Hydro-Electric Plant: Norsk Hydro (NOR)

By the outbreak of the Second World War, the Vemork hydro-electric plant had become the most important, if not the only significant source of heavy water in Europe. With the warning provided to President Roosevelt by Einstein, it was inevitable that Rjukan would become of pressing interest to the Allies and the Vemork plant a target for destruction. The problem was how to undertake such an exercise in winter, at a location so distant from Great Britain— in the remote interior of Norway. The answer was seen to lie with the recently created SOE and a force of British Army commandos.

Preparation for the Rjukan Raid

The assault on the Vemork plant was originally conceived as a joint SOE-Commando exercise, in which an initial party of Norwegian agents would prepare the ground, including finding a suitable landing field. This would be followed by the insertion of a glider-borne force of commandos who would carry out the attack.

In the event, while the preparatory stage, codenamed Operation Grouse, was completed successfully, the second stage was a complete disaster. The entire commando force, codenamed Freshman, was lost when both its components crashed on the western coast of Norway, far away from the intended landing spots on the Hardanger plateau to the west of Oslo. Any survivors were summarily executed.

The initial four-man party of Norwegians was thus left to fend for itself for several months, their only food what could be found locally, in the middle of one of the more hostile parts of Europe in midwinter. Nevertheless, the Grouse radio operator, Knut Haugland, was able to retain contact with the SOE base station in England with an early SOE-designed radio, the Type 3 Mark 1 (also known as the Type B Mark 1). This radio is now on display at the Norwegian Resistance Museum in Oslo, which for many years was curated by Haugland in his civilian life.

The loss of Freshman forced a fundamental reassessment of the means to attack the Vemork plant, and the decision was made to send in a further group of SOE-trained Norwegian agent saboteurs. Again this attack was to be supported by the four Norwegians already in place. To signify the change of plan, this group was renamed Swallow. The additional party of saboteurs was referred to by the codename Gunnerside.

The Swallow party—leader Jens Poulsson, radio operator Knut Haugland, Claus Helberg and Arne Kjelstrup—was joined by the new leader of the whole operation, Joachim Ronneberg, and the remainder of Gunnerside: Knut Haukelid, Kaspar Idland, Frederik Kayser, Berger Stromsheim and Hans Storhaug.

The raid that followed has been portrayed in the rather inaccurate but well-known Hollywood extravaganza, *The Heroes of Telemark*, with Kirk Douglas portraying a character representing Ronneberg in a fashion that caused considerable offence in Norway. What is well portrayed in the epic is the Norwegian landscape near Rjukan and the Norsk-Hydro plant at Vemork.

For a realistic portrayal of the activities of the saboteurs in this remarkable raid, a joint Norwegian–French production of 1948 can be referred to. The film, available with some difficulty over the Internet, is titled *Kampen om Tungtvannet* ('The Battle for Heavy Water'), and describes in appropriately matter-of-fact and quite undramatised fashion the landing of the Grouse party on the Hardanger plateau and their subsequent wait for the arrival of the Gunnerside party. It also shows an accurate portrayal of the attack on the Vemork plant and the demolition of the electrolysis cells in the basement of the multi-storey building, which was followed by the escape of the saboteurs without a shot being fired.

As is apparent from both films, the Vemork plant was exceptionally difficult to approach due to natural topographic features. Set on a projecting spur in the upper part of the valley in which Rjukan is located, at first sight it appears as inaccessible as a medieval castle on a mountaintop.

View from the bridge to Vemork (PRJ)

Despite this, the saboteurs were able to descend 200 metres into the steep-sided gorge and climb up the far side to reach the plant. They were able to access the plant and carry out the planned demolition without disturbing the German guards—who may have assumed the muted explosion was from a land mine set off by the weight of freshly fallen snow.

The raid elicited the wry admiration of the Germans when they came to contemplate the damage to the electrolytic cells in the basement. As General von Falkenhorst commented, '[It was] the most splendid coup that I have seen in war'. This did not stop him from also referring to the saboteurs as 'British gangsters'. Fortunately for the local population, the fact that the saboteurs were actually Norwegian was unknown—the British Sten gun that was deliberately left behind was intended to lead to the wrong assumption.

Communications and Coordination in the Rjukan Raid

While the Rjukan raid is one of the epic events of the Second World War, for persons interested in radio it is the activities of the Grouse radio operator, Knut Haugland, that are of particular interest. As made clear in his personal file from the National Archives at Kew in Great Britain, recently made available on the Internet, he was a quite remarkable person who was to be recipient of a number of Norwegian, British and French awards including a British Military Medal and the Distinguished Service Order.

Type 3 Mark 1 transmitter receiver (PRJ)

Haugland was the epitome of the 'quiet achiever'. Described as 'Keen, efficient and intelligent but a bit on the retiring side ... he impressed us with his quick thinking and unfailing attention to detail', it is apparent that here was an ideal officer to take part in a desperately dangerous task as well as someone extremely well qualified to provide the vital radio link to England and the SOE coordinators.

The National Archives has placed Haugland's own report of the Rjukan raid in the public domain and this makes particularly interesting reading. The report makes very clear the sort of problems that had to be dealt with in establishing shortwave contact with Great Britain under the extremely adverse conditions that applied during the winter of 1942.

The radio transmitter receiver supplied to Operation Grouse was a Type 3 Mark 1. As set out in the clandestine radio 'bible', Volume 4 of *Wireless for the Warrior* by Louis Meulstee, this is a somewhat higher powered radio than the Paraset, but certainly no lightweight. Although contained in a suitcase like the Paraset, it is an arm-destroying 19 kilograms.

According to Meulstee, the Type 3 Mark 1 was capable of producing between 9 and 18 watts of RF in CW mode, depending on the form of power supply in use. The transmitter used a 6L6G as the RF generator which was driven by a crystal. The receiver was a superhet rather than the regenerative detector employed in the Paraset, and thus would have been far easier to operate. Indeed, Haugland's report makes it clear that as a professional operator he had nothing adverse to say about the radio. What he did complain about was the overall weight of the apparatus. This consisted of the radio, an accumulator, a Eureka aircraft communications set and a hand generator—approximately 90 kilograms in all, a major burden in the context of the other essential equipment the saboteurs needed to carry through the freshly fallen snow.

Initial attempts to contact SOE in Great Britain were unsuccessful and Haugland refers to some of the probable reasons. Not least was the problem of damp penetration into a radio that was not hermetically sealed and which needed to go through a cycle of heating and cooling. This produced internal short circuits and blown fuses, putting the radio out of action for a considerable period. Operation Grouse departed for Norway on 18 October 1942 and Haugland was not able to achieve contact until 9 November. Unsurprisingly the receiving station was suspicious of this delayed signal, and it was not until a prearranged trick question was successfully answered that the contact was confirmed.

The message read:

> Happy landings inspite of boulders everywhere STOP Sorry to keep you waiting for message STOP. Snow storm fog forced us to go down valley STOP. 4 feet of snow made it impossible with heavy equipment to cross mountains STOP.

In his later report Haugland referred ruefully to the boulder field and observed that it was a miracle that no one had been killed during the parachute landing.

Quite apart from the problems of damp and the shorting of internal components, expecting to send a signal as far as England with a physically minimal antenna and low power was exceptionally optimistic, and something that the planners of the raid should have made better provision for. As described by the participants, elevating an antenna on the open snowfield was extremely difficult. Why some form of telescopic vertical mast was not supplied to meet exactly this situation is somewhat baffling— but the benefits of hindsight, as ever, can be seen as irrelevant. With only a tent to protect the radio and its operator from the elements, erecting a useful antenna with reasonable radiating characteristics was probably impossible, although skis were lashed together to achieve some degree of elevation.

Even taking into account Haugland's inventive abilities with the only source of electric power available, a motor vehicle type battery operating under winter conditions, it is probable that the RF output of the Type 3 Mark 1 was at the lowest level, at perhaps 8 to 10 watts, as developed from the available vibrator type high voltage supply. As the antenna erected on a couple of skis could not have been more than 2 or 3 metres above the

Agent operating a Type 3 Mark 1 transceiver in the snow with wire aerial (INT)

snow surface, radiated power would have been far less than the power available from a properly matched quarter-wave vertical.*

The antenna problem was eventually overcome when the Grouse party was fortunate enough to find an empty hut on the plateau and erect a more substantial antenna, made from a mast, attached to the roof. At last Haugland was able to make contact with the listeners at Whaddon Hall.

Following the Freshman disaster, the Grouse party remained on the plateau, making do with ever dwindling rations until forced to eat moss and lichen scraped from the rocks. It was several weeks before their leader, Poulsson, was able to track and shoot a reindeer, saving the saboteurs from complete starvation. It was later reported that everything but the testicles and the hooves were consumed.

The Gunnerside party arrived on 16 February 1943, and on 10 March 1943 Haugland was able to send the following message to SOE:

High concentration installation at Vemork completely destroyed on the night of the 27th—28th STOP Gunnerside has gone to Sweden STOP Greetings STOP.

---

* In this context, an assessment of the kind undertaken by Brian Austin in the pages of *Radio Bygones* (Vol. 120, 2009) would be useful, coupled with an assessment of the state of the ionosphere in late 1942 and the path to England across the North Sea.

Qualified Success

While the raid itself was a complete success, the Germans had been prepared for such an eventuality and had stockpiled a substantial volume of heavy water. Moreover, they were able to repair the electrolytic cell installation and resume production by April 1943. American pressure now led to the carrying out of a bombing attack. Although abortive, this appeared to convince the occupation forces that a strategic withdrawal was appropriate, and steps were taken to evacuate the heavy water supplies to Germany by rail.

Moving such a load to the coast involved traversing the elongated Tinnsjo Lake by means of the ferry SF *Hydro*. Knut Haukelid, one of the six Gunnerside saboteurs, had elected to remain in the vicinity of Rjukan. As the only available SOE-trained saboteur, he was instructed to intercept and destroy the train-load of heavy water. This was done by setting off a demolition charge on the *Hydro* as it was crossing the deepest part of the lake, which had the unfortunate although anticipated consequence of killing a number of innocent civilian passengers.

Lake Tinnsjo ferry and rail terminal (PRJ)

Rjukan Today
Although Rjukan is an interesting destination for the dedicated student of communications history, in one respect at least the hydro-electric plant has seen a dramatic change. Some years ago the multi-storey building that sat in front of the original hydro-electric plant was demolished. In 2010 the site had a very different appearance to that seen by the Norwegian agents of 1942 (compare this illustration with the picture on page 146). Nevertheless, the difficulties of accessing remain perfectly obvious, in particular the depth of the gorge over which the access road is suspended.

The Vemork power station, 2010 (PRJ)

Rjukan lies a considerable distance from Oslo and accessing it by public transport is somewhat complicated, involving several changes of buses. There is no direct bus access so the historically inclined traveller is forced to approach the station buildings on foot and up a steep incline from the suspension bridge across the ravine.

Clandestine Operations: Coastwatchers and Force 137 (Z Force)
In Europe a secret agent's major problem was being able to converse in the local language, but in the Southeast Asian and Pacific Rim countries nearest to Australia complexion and physiognomy presented the greatest

difficulty. It was impossible to disguise the white skin of most Australians. It was local populations that undertook the clandestine activities required to respond to the advancing Japanese Army after the start of 1942.

During the 1930s, an informal coastwatching service had been established in the islands to the north of Australia. Largely manned by civilian operatives and missionaries, this service, providing reports on shipping and air movements, became increasingly more important and dangerous as the Japanese Army controlled more and more of the territory to the north of the Australian mainland. The coastwatchers used the civilian-designed Teleradio manufactured by AWA.

Coastwatcher receiving message from Australia with AWA Teleradio 3B (AWA)

From its inception in 1935 with Teleradio 1, AWA developed this apparatus progressively until by 1940 it had reached the designation 3A. The radio was contained in two separate steel boxes, the first containing the transmitter and the second the superhet shortwave receiver. Soon afterwards, in April 1940, AWA produced an upgraded version of this radio, designated 3B, and in this apparatus, the receiver covered a range of 200 kilohertz to 30 megahertz in five switched bands. The 3BZ receiver and transmitter soon followed, with various refinements, with the station able to be powered by a conventional mains AC supply or batteries. Even more importantly, all this apparatus was provided in sealed steel cabinets with a tropicalised varnish applied to components internally to protect them from damp and fungus attack.

Operation Jaywick

The Inter-Allied Services Department (IASD), an Allied military intelligence unit, established in Australia in March 1942 to coordinate intelligence in the South-West Pacific theatre, was modelled on the British Special Operations Executive (SOE) in London. A raiding commando unit, known as Z Special Unit, was soon established. Typical of its actions was Operation Jaywick, undertaken with the aid of a Japanese fishing vessel, renamed the *Krait* after a particularly venomous snake that lives in the jungles of Indonesia and Malaya—entirely appropriate given the nature of the attack that was launched on Japanese vessels lying at anchor in the roads and harbour at Singapore.

The attack was carried out with folboats, collapsible canvas two-man canoes, carried on the *Krait* and launched some kilometres away from the intended target. The expedition under its commander, Captain Ivan Lyon, an officer in the Gordon Highlanders, was a huge success and seven Japanese vessels were destroyed or seriously damaged by magnetically attached underwater limpet mines. The men of the *Krait* and the invading canoeists were able to escape to Australia.

Horace (Horrie) Young, radio operator for Operation Jaywick, who died in 2011, a year or two earlier told the story of the radio installation he used. It involved a transmitter and receiver pair originally developed for the Royal Australian Airforce by AWA to provide a general-purpose high-frequency and medium-frequency capability. On MF, the coverage was from 140 kilohertz to 2 megahertz, and on HF, from 2 to 20 megahertz. To cover this extended range, the receiver was constructed in two modules, set side by side in the same housing, while the transmitter occupied a single case a little bigger than the receiver housing. In order to match the transmitter receiver to a variety of antennas, a substantial tuning unit was provided in a case approximately the same size as the transmitter. The three elements of the radio station can be seen in the illustration overleaf.

The transmitter was capable of generating between 25 watts at 20 megahertz and 45 watts at 2 megahertz, but on the *Krait* these outputs were attenuated by a very inefficient antenna system. Initially this had involved a wire from the bow of the vessel over the main mast to the rear of the boat, but when the mast was removed to reduce the silhouette on

Receiver AR8 with transmitter AT5 and antenna tuning unit by AWA (PRJ)

the horizon, another means to radiate the RF signal had to be found. This was achieved by running a long wire around the gunwale of the vessel which meant that it was only a few feet above the water. Later, to improve the radiation pattern and efficiency, Young organised a sheet of copper to be attached to the underside of the wooden hull. The wonder is that the signal got out at all, and certainly to be heard in Darwin from the Java Sea was quite a feat.

As it turned out, there was very little transmission work as the operation was conducted in radio silence—apart from a single message advising of the success of the expedition once the *Krait* was well away from Singapore en route back to Western Australia.

Sparrow Force on Timor and Winnie the War Winner

Shortly after the Japanese attack on Pearl Harbor and the fall of Singapore, Japanese forces drove south-east along the Indonesian archipelago, finally invading the island of Timor. The campaign very soon overpowered the limited Dutch and Portuguese colonial forces. However, just a few days after the attack on Hawaii, an Australian Army contingent designated Sparrow Force had been landed on Timor to support the Dutch and Portuguese.

As the Japanese assault continued, Sparrow Force divided into two components, one located in the vicinity of Dili on the north coast in the Portuguese zone, the other in the area around Koepang in the Dutch zone at the south-west corner of the island. In the face of further enemy advances, both elements of Sparrow Force retreated into the heavily forested hinterland where for a considerable period they were effectively cut off. In the retreat, much equipment had to be abandoned—even the radio sets. These were WS No. 101s and WS No. 109s that in operation had proved to be seriously deficient, being incapable of contacting Darwin in daylight hours, having insufficient power as well as being assigned inappropriate frequencies on which to operate, at around 2 megahertz.

From their hinterland locations, Sparrow Force commenced a highly dangerous form of guerrilla warfare to harass the invaders, initially assisted by the local Timorese population. Conducting such a campaign without adequate supplies of food and ammunition was a prescription for disaster, and contact with mainland Australia became essential.

Strenuous efforts to make contact with Australia were made using a WS No. 101 transmitter receiver set retrieved from an area previously captured by the Japanese. Unfortunately, the set's low power ensured the failure of these attempts, and a means of increasing the power was looked for. A successful campaign of scavenging found the parts required to construct a radio frequency amplifier involving two type 4307A tubes (equivalent to the better known tube, the 807). The result was a piece of apparatus that is now housed in the Australian War Memorial in Canberra and known as Winnie the War Winner.

As noted by Captain George Parker, who played a significant part in the construction of this radio, the display in Canberra is deficient in that the front end of the apparatus, a WS No. 101, does not appear. Parker comments that three versions were constructed before the amplifier was able to produce a signal that could be heard in Darwin. Again this related to the problem of operating on frequencies that were too low to allow totally effective contact to be made, except at night time.

As revealed in papers in the Australian War Memorial Archives, the first, unsuccessful version of Winnie involved a pair of 6F6 valves driven directly by a crystal on a frequency of 4.547 megahertz. Later, this initial circuit was developed into a power amplifier to be used in conjunction with a WS No. 101 and feeding into a Zepp antenna.

This crude but successful item was assembled on a steel petrol can with wiring joined together using scrap solder from scavenged radio items. The transmitter used to drive this amplifier, a WS No. 101, was able to transmit a continuous-wave Morse code signal powerful enough to be heard in Australia using skywave transmission. When the signal was first

Winnie in operation with Jack Sargent on the key (RAS)

received, it elicited a great deal of scepticism as Sparrow Force had been out of contact with Australia for a considerable time. Only after a number of special questions concerning things that could only be known by genuine members of Sparrow Force were responded to correctly were the operators in Australia convinced that they were not dealing with the Japanese Army in disguise.

Several months later Sparrow Force was removed from Timor by sea and returned to Australia, having achieved a memorable feat in the annals of military and amateur radio. One of the principal constructors of Winnie was radio amateur Max Loveless. Others involved in the construction of the amplifier were Captain George Parker (8th Division Signals), who provided technical knowledge, Corporal John Sargent, John Donovan and Keith Richards.

The 6F6 valves initially used in the amplifier were later replaced with two type 4307A valves. These had been used in the WS No. 109, but in that radio only one was operated as the output valve. The other two were used in the crystal oscillator and driver stages and resulted in considerably less power output than was available from the pair operating in push–pull, as they were included in the Winnie amplifier. The simplicity of the circuit is apparent in the schematic taken from Captain Parker's original sketch, lodged in the Australian War Memorial Archives.

Winnie the War Winner: RF amplifier in the Australian War Memorial, Canberra (PRJ)

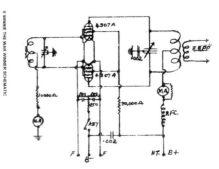

RF amplifier schematic (PRJ). Larger version p. 324.

Drive transmitter receiver—WS No. 101 (PRJ)

10 Technological Change: Valves, Miniaturisation and New Circuitry

In the inter-war period, the technology of communication underwent a dramatic change, initially induced by the advent of public broadcasting and then as a response to the advancing shadows of a new conflict. In this period television made an appearance, initially in a Heath Robinson version of spinning disks and glass lenses put together with string and wire. Later, television was developed in an electronic form that Alan Campbell-Swinton would have immediately understood, as it used electron beam scanning as he had suggested in 1912.

Changing Valve Configurations

The most conspicuous changes in radio between 1920 and 1939 were the developments in valve technology. The fragile Edison lamp-like devices of the 1914-18 war years were progressively replaced by a robust form of glass bulb construction that became ever more compact. The initial onion-shaped bulb exhibited in the French R valve gave way to a round-topped shape in the 1920s that during the early 1930s became more tubular in form. The electrode construction of the valve was progressively reduced in size and the external glass enclosure was able to be shrunk to enclose less and less unfilled vacuum space. This progression is best appreciated in the accompanying illustration.

The significant counterpoint to the physical contraction of the valve was the increase in the number of electrodes and grid structures between the anode and the cathode. The simple three-electrode configuration of the De Forest triode in less than ten years grew into the multi-element structures found in frequency converter valves, and in triode, hexode and pentagrid converter valves. The same configuration was typical of power amplifier

Technological Change: Valves, Miniaturisation and New Circuitry

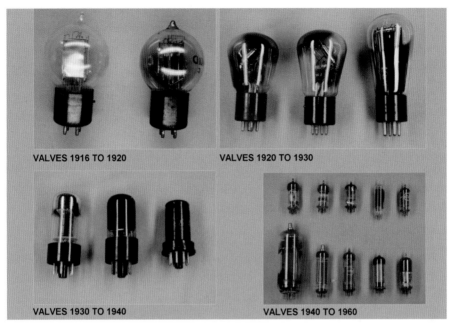

The changing valve—1916 to 1960

valves, where focusing of the electron stream from cathode to anode was also introduced. In these so-called 'beam tubes', increased efficiency and greater power was achieved without overstressing the tube in an electrical sense. As well known as any such beam power tube was the ubiquitous 6V6, which was used in many domestic receivers during the 1930s to provide substantial audio output.

Regeneration to the Superheterodyne

At the end of the First World War, the superheterodyne circuit of Edwin Armstrong and Lucien Levy had started to make an appearance in radio receivers. In the United States, as the broadcast bands became ever more crowded with competing radio stations, the need for improved selectivity generated consumer demand for change.

While the regenerative receiver was sufficiently sensitive to operate in the metropolitan context, its capacity to break into self-oscillation made it extremely unpopular with neighbouring listeners who found the heterodyne whistle unacceptably intrusive. The initial answer was the straight set without regeneration, but until a method of overcoming internally generated feedback was introduced the large number of tubes required to produce an acceptable level of audio in a loudspeaker was a problem. Neutralisation of individual stages, as incorporated in the neutrodyne receiver, proved popular for a while with the public, but the multiplicity of turning dials was a massive hindrance to acceptance by the listeners. What was required was a single knob tuning system, and in the superhet lay the answer.

By incorporating ganged tuning of successive RF and mixer stages with fixed frequency intermediate stages of amplification, the superhet satisfied the need for a sensitive, selective and high gain radio that was simple to operate. A single knob to control the tuning and another to control volume was complemented by a wave-change knob when listening to shortwave stations became increasingly popular during the 1930s.

Despite the superior performance of the multi-stage superhet receiver, the increased number of valves was a costly solution. As a means of reducing the cost, increasing the number of valve assemblies within each glass bulb was also used. With this arrangement, the number of valve envelopes could be reduced as compared with a conventional radio using single discrete valves.

Crystal Control

In the relatively uncrowded airwaves of the First World War and immediately afterwards, broad and unstable radio frequency signals were not a major problem and spark was able to exist for quite a while. As spectrum space became progressively more crowded, unstable signals were seen as unacceptable and self-excited oscillators in particular developed a very well-deserved and progressively bad reputation until ultimately they were banned.

The crystal oscillator could be made extremely stable as compared with a self-excited device, particularly with the application of sophisticated techniques involving thermal control and feedback. The inevitable problem of crystal control was that each crystal was tied to a particular frequency, thus to cover a number of frequencies an equal number of crystals was required. During the Second World War, crystals were assembled into banks of a dozen or more with a rotary switch to select the relevant pair, one for the transmitter, the other for the receiver. Later, the superhet approach was applied to transceivers, with a single crystal providing the control of both transmit and receive frequencies. This was introduced in the British WS No. 62 at the very end of the war and was also used in the Australian version of the British WS No. 22 (the WS No. 122).

The only problem with this approach was that operators had to remember that the transmission crystal for a particular frequency of reception had to have the value of the intermediate frequency (IF) added to it to get the correct value. Initially this was the common 455 kilohertz IF frequency but the technique became more complex as other values came into use, particularly in VHF radios that operated between 40 and 80 megahertz and employed double conversion to reach the applicable IF.

The Transceiver Appears

Right up until the late 1930s, the two functions of transmit and receive were undertaken by separate apparatus. Even in the WS No. 101, developed just before the start of World War Two, the receiver and transmitter were still discrete devices although contained in a common steel box. From a military perspective, the problem with this arrangement was that keeping a communications group together on a single frequency invariably proved difficult, and particularly so with less capable operators. While facilities for achieving a single frequency of operation (netting) were introduced, this remained a fundamental problem and the simplicity of operation of the BF trench set of World War One was clearly desirable.

The answer was to control the transmitter frequency with the receiver local oscillator mixed locally to the transmit frequency. Where crystal control was required, this would involve the use of a crystal with a frequency separated from the transmission frequency by the amount of the intermediate frequency. The result of this circuit arrangement was the creation of a transmitter and receiver in which tuning was controlled by a single knob or with a single crystal. Tuning up the transmitter would still require two other sets of controls, one for input or tank tuning and the other for antenna tuning, together with a variable inductor for optimum RF energy transfer.

Complementary to this development was the realisation that weight and energy savings could be made by making valves perform different tasks, depending on whether the device was transmitting or receiving. A most impressive early example of this technique was introduced in the Handie-Talkie, in which an elongated sliding pressel switch, running along the length of the radio, controlled the functions of various valves and components. This radio also incorporated the latest US miniature valves, the 1R5, 1T4, 1S5, 3S4 series, which drew minimal filament current from a dry battery pack which also contained a high-voltage section. The battery pack slid into a compartment that lay parallel to the main chassis of the radio and, considering the period, allowed the creation of an impressively compact communications device. However, being crystal controlled, and with a restricted space in which to include these devices, the BC-611 was restricted to a single frequency of operation. To change frequency required the installation of a different pair of crystals for receive and transmit, as well as coil boxes.

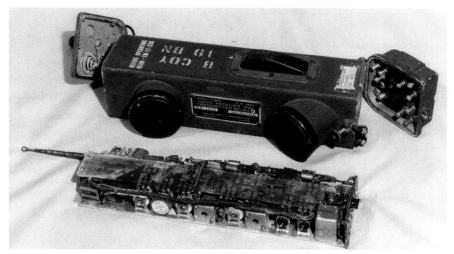

BC-611 Handie-Talkie open to show main chassis and slide switch (PRJ)

Evolving Military Radio

From the end of the First World War to the start of the Second, as the apparatus of radio developed—especially valve technology—the complexion of transmitters and receivers also changed, which could be seen in the increasing complexity of circuits. These changes are readily apparent from the schematic diagrams in the operating handbooks prepared by the military for the instruction of operators in the field and servicing personnel away from the battle.

Wireless Set Type 'C': the Cork set

Characteristic of the early 1920s, and reflective of apparatus developed during the First World War are the arrangements of the Cork set discussed in Chapter 6. This radio was provided in two discrete wooden boxes, one containing a 4-valve 'straight' set with reaction and the other a transmitter that employed a single, relatively high power valve.

Technological Change: Valves, Miniaturisation and New Circuitry

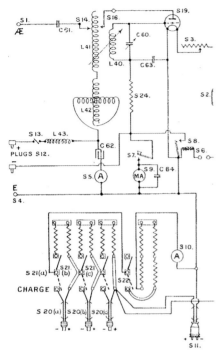

Cork set transmitter (KMM). Larger version p. 325.

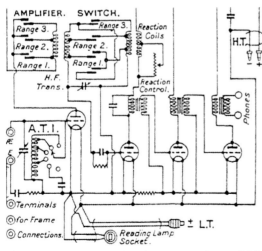

Cork set receiver (KMM). Larger version p. 324.

The single valve in the transmitter was an AT50 configured as a tuned-plate tuned-grid radio frequency oscillator and amplifier, referred to as a TPTG set. As radio amateurs of the 1930s were to discover, this was an inherently unstable and temperamental means of producing a continuous wave (CW) signal, and was later completely displaced by crystal-controlled oscillators or more stable configurations of oscillators in separate stages before power amplification.

The receiver was a so-called 'straight' set which employed regeneration in the second stage detector and was preceded by a tuned radio frequency front end. The detector stage incorporated a grid leak as part of the rectification circuit, and was followed by two stages of audio frequency amplification, incorporating audio frequency transformers with the output sent to headphones.

The receiver design was consistent with that employed in broadcast receivers available to the public in the 1920s and early 1930s, but unlike broadcast receivers this radio was intended to operate with the regeneration control set so that Morse code sent via CW could be resolved by beating the incoming signal against the oscillating detector. This configuration was referred to as an auto-dyne mode and, like the TPTG transmitter, was to be made obsolete by the progressive introduction of the superhet receiver circuit, with its superior sensitivity and selectivity and general ease of operation. However, for the detection of CW, a superhet required an oscillator to provide a signal that was mixed with the intermediate frequency to produce an audio frequency resultant. This oscillator was known as a beat frequency oscillator or BFO.

The four valves used in the receiver were designated as type AR3 and were conventional 4-pin triodes from the early 1920s. A peculiar feature of the use of these valves in the Cork set is that they were carried on the face of the receiver and were connected to the internal circuit via connector blocks. This allowed them to be mounted parallel to the front panel with the connector block set at right angles to the direction of the pins. Apart from this, the valves are extremely rare, which is rather discouraging to the impecunious collector.

Wireless Set No. 101

The WS No. 101 was based directly on the WS No. 1 introduced into the British Army in 1933. The WS No. 101 was supplied to the Australian Army in the late 1930s. The schematic indicates more than a passing resemblance to its predecessor, the Cork set. It was configured as two separate units, one containing the transmitter and the other containing the receiver.

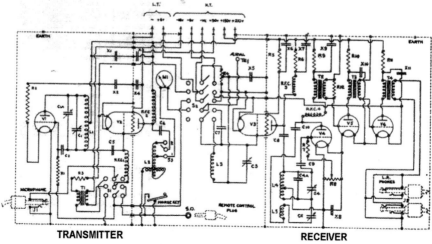

Wireless Set No. 101 (AMP). Larger version p. 326.

Technological Change: Valves, Miniaturisation and New Circuitry

As for the Cork set, a separate power supply was provided, although in this instance it was in a steel box of the same form of construction as the box containing the transmitter and receiver.

The WS No. 101 had a straight receiver involving an RF front end feeding into a regenerative detector as the second stage. The audio output of this stage was then amplified by two stages which fed into earphones. The transmitter, unlike that of the earlier Cork set, employed an initial stage as the tuned master oscillator which fed into an amplification and RF output stage. This was generally referred to as a MOPA configuration. The output stage was also able to be grid-modulated with an audio signal from a carbon microphone, so the radio was able to send and receive voice messages as well as send Morse code with CW.

Wireless Set No. 109

The WS No. 109, used in the early days of the Second World War by Australian forces in North Africa, was related to the WS No. 9 designed for the British Army and introduced in 1939. Designed in Australia by Standard Telephones & Cables, the WS No. 109 represented a quantum leap from the relatively primitive WS No. 101 and was significantly different to the British parent radio, WS No. 9. Radio design as it was developing in the United States may have influenced the approach taken in the Australian version.

As the schematics for WS No. 109 indicate, here was a fully featured superhet receiver as well as a crystal-controlled transmitter which included an audio modulation stage. Moreover, the WS No. 109 was considerably lighter than the WS No. 9, although it produced a lower output RF signal at between 5 and 7 watts. Although provision for crystal control of the master oscillator was available in this radio, it appears that it was not very often

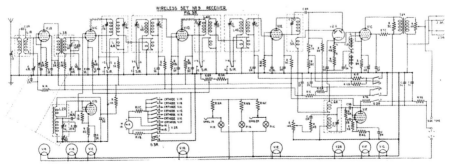

Wireless Set No. 109: receiver (AMP). Larger version p. 327.

employed. As with the WS No. 9, in initial versions of the WS No. 109 there was no hermetic sealing of the chassis and the radio was susceptible to damp ingress and fungal attack, a problem rectified in later versions.

Although contained in a single steel case with hinged front cover, the discrete nature of receiver and transmitter ensured that bringing them onto a common frequency remained a problem, and netting remained a skill that army signallers had to acquire through training and experience.

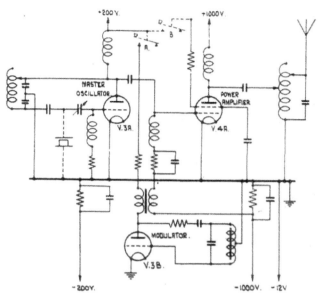

Wireless Set No. 109: transmitter (AMP). Larger version p. 327.

Wireless Set No. 11

Unlike earlier military field radios, the WS No. 11 would now be called a transceiver—that is, a radio in which the transmitter and receiver are contained in a single steel box and whose components have two functions, depending on whether the radio is sending or receiving. In the WS No. 11, three of the valves are used both for transmission and reception.

For field operations, the most important advance in this radio was the use of the local oscillator to determine the frequency of reception of the receiver and an identical frequency of operation for the transmitter. This again would become characteristic of all later transceivers, both for military and civilian use. In the military context, the enormous advantage of this arrangement was that the problem of netting to a common frequency in dispersed troop locations was overcome. Finding other operators in a net became a simple process of correctly locating the master control signal, when automatically the local transmitter would be tuned to the same frequency.

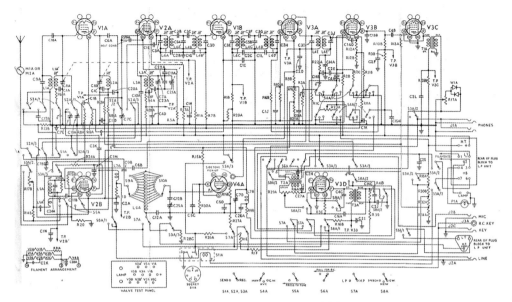

Wireless Set No. 11: Australian (AMP). Larger version p. 328.

Wireless Set No. 19 (Aust.)

The disadvantages of the WS No. 11 as developed in Australia—excessive weight and susceptibility to moisture—fairly quickly led to the adoption of one of the best-known field- and vehicle-borne radios of the war years, the WS No. 19. This radio was not available for distribution to Australia, so a locally designed version which was plug-compatible with the British radio was produced and introduced into the Australian Army as the WS No. 19 Mk 2 (Aust.) (see schematic overleaf). The development of this radio was considerably expedited by the use of local versions of American octal-based valves, available from Amalgamated Wireless Valve in Sydney.

Wireless Set No. 19 Mk 2 (Aust.) was considerably lighter than the WS No. 11, able to be man-carried, and was coated internally with varnish to prevent fungus attack as well as having a rubber gasket seal around the front panel. Coupled with sealing around the control knobs and other items that penetrated the front panel, this achieved a substantial improvement in tropical operation and contributed to its general usefulness.

The WS No. 19 Mk 2 (Aust.) retained the advantages of a common tuning control for both transmission and reception, which almost entirely overcame the problem of maintaining a net for a number of dispersed radio stations, either man-carried or in tanks or other vehicles such as the Jeep.

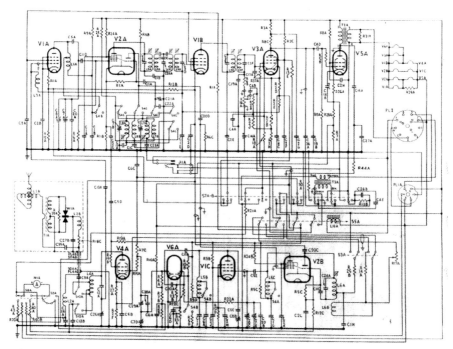

Wireless Set No. 19 (Aust.) (AMP). Larger version p. 329.

Wireless Set No. 62

In 1942 the WS No. 22 was supplied to the British Army but a number of failings, including low power output, meant that it not an outstanding success. At the disastrous Battle of Arnhem, the failure of British glider- and parachute-borne troops to secure the bridge across the Rhine related in part to problems with communications with the WS No. 22 and other, lighter backpack radios. Reliance on whip antennas rather than substantial wire aerials, as well as the supply of incorrect crystals, contributed to a major breakdown in communications in the face of the unexpected scale of the German Army's reaction.

As a result, alternative approaches were taken in the design of a successor to the WS No. 22. In Australia, a locally designed version was developed as WS No. 122, with crystal control available. In Great Britain, Pye Telecommunication produced an upgraded design designated WS No. 62. This radio was issued in 1945 and remained in service for a considerable period. The WS No. 62 was also made under licence in

Australia and remained in service until the 1960s in Citizen Military Forces.

The WS No. 62, being directly derivative of the British No. 19 and No. 22 sets, shared a number of their characteristics. The radio was a true transceiver housed in a single metal box, in this instance incorporating an aluminium front panel rather than steel. The power supply was incorporated within the same metal box, and extensive use of rubber gaskets allowed temporary water immersion without damage. However, the outstanding feature of this new radio was its reduction in weight, being around half the weight of the WS No. 19. At last the British and Australian armies had access to a truly portable transceiver with ease of operation and sufficient output power. However, the WS No. 62 operated on the HF radio frequency spectrum between 1.6 and 10 megahertz, with the inherent problem of low radiation efficiency associated with steel-rod whip antennas rather than wire aerials, with dimensions close to natural quarter-wavelengths of the frequency in use. By comparison, by the end of World War Two, US radios for short-range battlefield communication were operating in the very high frequency range, between 40 and 80 megahertz, with conspicuously greater success than British radios.

The schematic for the WS No. 62 reveals its general similarity to its predecessor, the WS No. 19. This includes the generation of the transmitter frequency by mixing with the receiver local oscillator and allowing a single tuning knob for both receiver and transmitter.

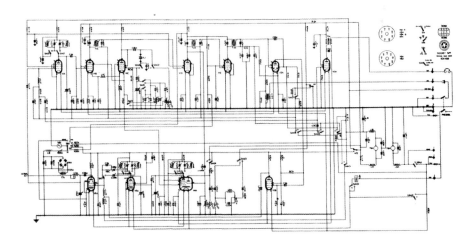

Wireless Set No. 62 (AMP). Larger version p. 330.

US Infantry Radio BC-611: The Handie-Talkie

As earlier noted, in the latter part of World War Two, the complexion of Australian Army radio began an inexorable shift towards US-mediated design. The very useful, although somewhat limited, Handie-Talkie or BC-611 hand-held transceiver was introduced for use by front-line infantry troops.

As the accompanying schematic reveals, this was a relatively simple transceiver whose compact form resulted from a combination of miniature valves and shared functions of common components, achieved by the inclusion of an elongated sliding switch that changed the function of components depending on whether the device was receiving or transmitting. From the schematic it can be seen that in the receive mode, five miniature valves function as a superhet receiver, whereas in the transmit mode, three of these same valves are used as voice modulator, crystal-controlled oscillator and power amplifier.

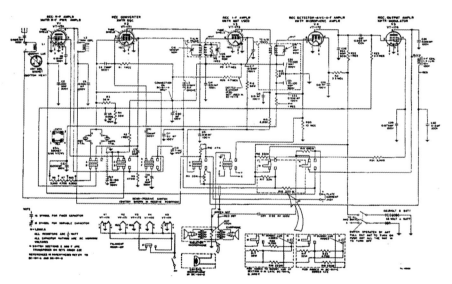

BC-611: the Handie-Talkie (AMP). Larger version p. 331.

US Infantry Radio BC-1000 VHF FM

The introduction of VHF in infantry communications in the US Army in the early part of the Second World War is exemplified in the BC-1000, the transceiver that was influential in the later design of Australian Army radios. Quite apart from its use of the new miniature valves and its ability to operate at frequencies from 40 to 48 megacycles per second in 200 kilohertz steps, this radio used frequency modulation (FM), was constructed with a lightweight aluminium chassis and external container, and was designed to combat tropical conditions of high humidity and fungal attack. The high operating frequency coupled with a whip antenna of a length to match that frequency provided reliable communications over a distance of 5 kilometres from a transmitted power output of 500 milliwatts (mW). This capability was made apparent in the Market Garden campaign of September 1944, when US paratroops prepared part of the access route towards Arnhem. The success of the BC-1000 highlighted the failure of the WS No. 22 in the Battle of Arnhem. Frequency modulation at VHF was clearly the superior medium for battlefield short-range communications.

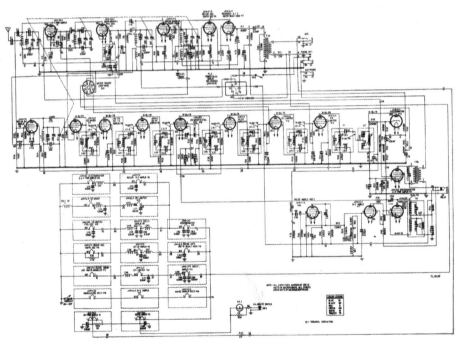

BC-1000 transceiver: the Walkie Talkie (AMP). Larger version p. 332.

Type A Mark 3 Clandestine Transceiver

Introduced in Great Britain in 1944 for the Special Operations Executive, the Type A Mark 3 transceiver was the most compact and useful of the various radios produced during the Second World War, although it appeared too late to have much impact on operations in Europe. As war surplus the Type A Mark 3 was exported to Australia in significant quantities and used for various civilian and quasi-military purposes after 1945. Considerable numbers were purchased by radio amateurs as a very satisfactory replacement for equipment surrendered to the government in 1939.

As the schematic reveals, the Type A Mark 3 was a transceiver in which valves served for both transmission and reception as controlled by the changeover switch and contained in a small steel box. The receiver was a 4-valve superhet in which the front end mixer valve was also used as the crystal-controlled oscillator for the transmitter. Morse code received as CW was detected by means of regeneration in the second detector and then supplied as audio to the last stage of amplification.

The transmitter consisted of two stages with the crystal oscillator stage driving a power output stage which, in fundamental mode, produced 5 watts of power. At the second harmonic of the crystal frequency, the power output stage generated 3.25 watts. The radio's frequency range was 3.2 to 9.00 megahertz.

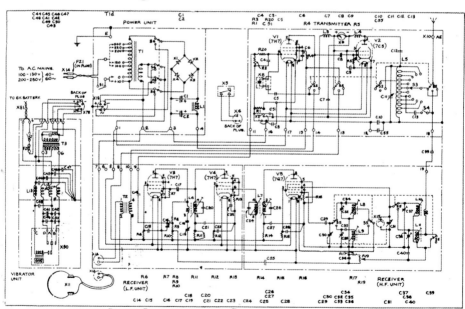

Type A Mark 3 clandestine transceiver (AMP). Larger version p. 333.

The Type A Mark 3 could be powered from the AC mains at either 240 or 120 volts, and also had a vibrator power supply which drew current from a 6 volt automobile battery. As the schematic indicates, the mains power was converted to high-voltage direct current to drive the valves by means of an autotransformer. As a result the chassis of the radio could be at full mains voltage, and for this reason a protective fibre box was contained within the outer steel casing. Switches and knobs, as well as the Morse key, were carefully insulated to keep the operator away from the potentially 'hot' chassis.

PROJECT 2: Paraset Replicas

The inspiration for the Paraset replicas was the book *Secret Warfare* by the French architect Pierre Lorain, which described the arms and techniques of the French Resistance and included a chapter on telecommunications apparatus, as well as describing the creation of coded messages. A series of elegant line drawings illustrated the various pieces of equipment provided to the French Resistance from Great Britain.

One of the radios used in France by SOE agents was the compact and relatively lightweight Paraset. This name derived from its anticipated use by operatives who were parachuted into France, although its official designation was the Whaddon Mark VII. This name related to its place of manufacture, undertaken by a small group of dedicated radio amateurs assembled by SOE for this unusual task.

Operation of the Paraset

Before describing the construction of the Paraset, reference to the operation of the circuit makes understanding what components are required somewhat easier. The accompanying schematic reveals a very simple transmitter and receiver combination.

The transmitter had a single stage consisting of a metal-cased 6V6. The transmitter was controlled by single crystals that were selected and inserted depending on the frequency required. These crystals needed to be quite electrically robust as they were driven hard. Modern miniature crystals are not suitable, as they are likely to overheat and crack in operation.

The receiver used two metal-cased valves, 6SK7, which operated as a regenerative detector followed by an audio amplifying stage which drove high-impedance earphones. The second stage was choke capacity coupled to the earphones; the choke had a value of 30 henrys. This component has proved to be particularly elusive for many constructors and the solution to providing it is discussed later.

Construction of the Wooden Box Replica

Olaf Reed Olsen's book *Two Eggs on My Plate*, and a visit to the Imperial War Museum in London, revealed that the first version of the Paraset was mounted in a wooden box. This information was sufficient to get the project started. The production of the replica was then undertaken in stages.

Stage 1: Collecting Background Information

The first step was to collect as much useful information as possible where, with access to the Internet the difficulty proved to be the quantity of material that was available rather than its scarcity. Winnowing this mass of information for the most relevant and unambiguous guidance was a time-consuming exercise that eventually led to the most useful site of Bengt Falkenberg, a Swedish radio amateur. Here all the details that are

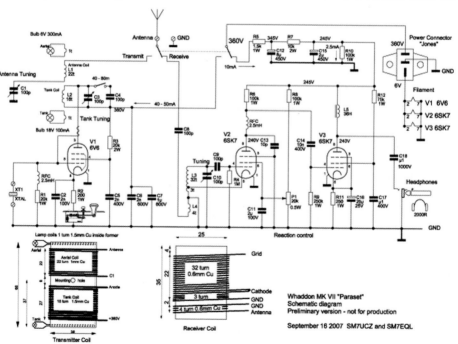

Schematic for the Paraset (SM7). Larger version p. 334.

required to build an accurate replica can be found. The specific site to visit is: http://www.sm7eql.se/paraset.

The subdirectory 'Download' reveals detailed dimensioned drawings of the metal top panel on which the components are mounted as well as the dimensions of the metal box version. Also to be found here are a circuit diagram and a component layout which allow a very accurate replica of the original Paraset transmitter/receiver to be made.

Stage 2: Collecting Main Components and Hardware
This stage proved to be quite difficult. The items required included the metal-cased valves and the components listed in the tables on pages 187 and 188. Very few of these items, or the necessary hardware, are now available from electronic components suppliers, so that access to historical radio societies is almost mandatory. The Historic Radio Society of Australia (HRSA) or an equivalent such as the British Vintage Wireless Society (BVWS) give access to many of the parts required for a successful replication project as well as a well-stocked personal 'junk box', while eBay can also be a useful resource.

Obtaining metal-cased valves proved difficult but the pages of the HRSA magazine *Radio Waves* led eventually to suppliers in Queensland and Paddington in New South Wales. The pie-wound RF chokes were also located through this resource.

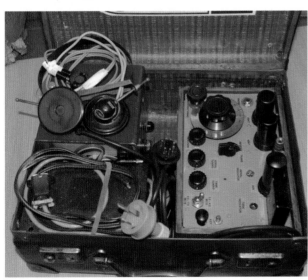

The wooden box Paraset replica (PRJ)

The personal junk box disgorged three 150 picofarad variable tuning capacitors with projecting shafts. Some surgery was required to achieve the necessary 100 picofarad value shown in the circuit diagram. This involved removing four plates per variable capacitor, using long-nose pliers to twist the plates out of their seating on the rotor arm and the stator rods.

The low frequency (LF) choke also proved elusive and eventually an audio frequency inter-stage coupling transformer from the early 1930s provided the solution. In this the two windings were connected in series to achieve the necessary 30 H inductance.

The miniature Morse key is quite unusual and here the choice was between using an unsuitable substitute or fabricating the device from scratch. The drawings at the Swedish site provided the incentive to undertake construction, assisted by the information that chloroform could make pieces of Perspex adhere to each other. The first Paraset replica included a full-depth brass arm but in the second brass strap was used. (See illustration opposite.)

Sheets of Perspex of the correct thickness came from some abandoned display shelving. This was duly cut up and glued together with the chloroform to create the chassis from which the brass arm of the key is suspended. The little key works surprisingly well and the dimensions follow the Swedish drawing below.

The spring steel clips, with which the metal-cased valves are stored in the lid of the metal-case version of the Paraset, were available at the local hardware store. Sold in sealed packs of three, they are used for retaining the handles of brooms in a storeroom. The addition of some heat-shrink tubing helps to prevent scratching the paint surface of the valves.

The Paraset makes use of a number of tag strips to wire up the smaller components. These are a standard item in the catalogue of a local electronic supplier, as are plastic (rather than metal) tuning indicator bulbs. Not completely authentic but physically identical to the real thing, and anyway well hidden in the forest of other components on the front panel.

Capacitors and resistors represented a minor problem due to the voltages and wattages involved in the circuit. High-value resistors with a power-handling capacity of 2 watts are no longer commonly available and,

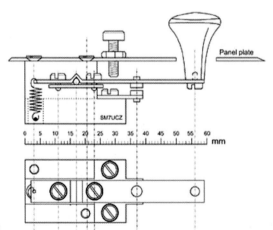

Morse key drawing (SM7)

in addition, contemporary resistors are generally much smaller than those made in the late 1930s. A number of suitable resistors, whose values were carefully measured before use, came from the junk box.

In the case of the electrolytic capacitors, many unfortunate experiences using old components in the past meant that new stock was used. Here the potential problem of subsequent fault-finding around a 'hot' chassis totally overrode the desire for complete authenticity.

Stage 3: Front Panel Metalwork and Assembly of Components

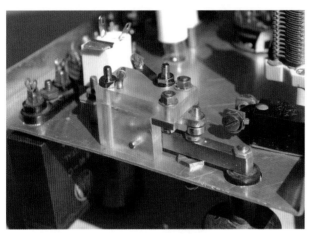

Morse key mounted on panel (PRJ)

The next step was the construction of the metalwork and assembly of the main components. All the components of the Paraset hang off the top panel, and their assembly requires metalworking tools, including a circular hole punch (if available) to make the openings for the three sockets for the metal valves. Other, rectangular openings required the use of a metal nibbler tool, metal files and a small hacksaw.

There were two common forms of enclosure for the Paraset components. The earliest versions were contained in a wooden box that could be carried in a small fibre suitcase, a common item during the war years. Later versions were enclosed in a metal box with a hinged lid in which the spring clips previously described were mounted to store the three valves.

To position the components accurately, the top panel schematic was printed out from the Falkenberg website and taped onto the aluminium top panel. In the second replica, the top panel was made of steel sheet. A spring-loaded centre punch was used it to mark out the centres of the various holes and components, which ensured accurate drilling using a manual drill press. Before the marking out, the two long edges of the metal sheet had to be bent up as flanges, which was done with a small pan-brake press.

Paraset under-panel construction (PRJ)

Wiring up the components could now be done. Note that one unpleasant feature of this radio is that one of the transmitter tank coil tuning capacitors is 'hot', making the radio potentially lethal. As the circuit diagram reveals, this capacitor floats at the voltage supplied to the transmitter valve anode at about 360 volts. Therefore it needs to be isolated from the chassis and precautions also need to be taken to ensure that the tuning knob does not provide a means of contact between the operator and the rotor spindle of the capacitor. This 'hot' variable capacitor was isolated from the chassis with two small ceramic stand-offs. Its spindle was covered with a small polythene tube, which runs from the underside of the knob, through the hole in the chassis and down to the shoulders of the capacitor below. The grub-screw in the knob is well recessed and was covered with sealing wax as an added precaution. Again it is appropriate to point out that the voltages available in this chassis are potentially lethal, so every care should be exercised both in construction and in later testing.

This radio was intended for use in France where, due to the high risk

PROJECT 2: Paraset Replicas

The Paraset wired up (PRJ)

of detection by the German radio intercept operators, agent-to-agent contact was specifically prohibited and facilities for setting the receiver accurately to the same frequency as the transmitter were not provided. This process is referred to as netting. It must have been assumed that the field agent would be listening to a powerful transmitter in Great Britain so that tuning the receiver would be relatively easy.

The transmitter, however, was always locked to a particular crystal frequency so that the British receiving station, with its sensitive and accurate receivers, could locate the clandestine transmission quite easily.

Operation of the replica in the radio amateur bands would be impossible without a means of operating with transmitter and receiver on the same frequency, so finding a method of netting was essential. Based on a suggestion from US radio enthusiast Paul

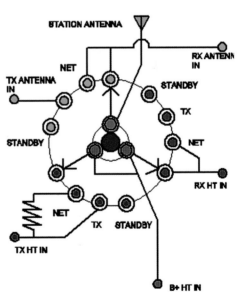

TX-RX changeover and net (PRJ)

Signorelli, an arrangement was devised in which a three-pole, four-position switch was used to set the radio to four different modes of operation. These were receive, standby, transmit and net.

In the net position, the receiver is still active and is coupled to the antenna so as not to de-tune the regenerative circuit. At the same time, a very low voltage is permitted to access the anode of the transmitting valve. In this mode, pressing the Morse key will generate a very low level CW signal which can be resolved by the receiver as a 'chirp' when the correct frequency is reached.

In addition to the functions of net, transmit and receive, a fourth position was included for standby. In this, the heaters of the valves are alive but no HT is applied to either receiver or transmitter. The advantage of this is that when a solid state rather than a valved power supply is used, the high voltage is not applied to the valves until the heaters are fully heated up, which may take approximately 30 seconds. This prevents the valves being damaged by a premature application of high voltage.

Another small but important modification of the circuit involved adding a couple of components to eliminate the 'key clicks' that occur in a radio frequency oscillator when the current flow is interrupted very suddenly, as happens when the Morse key is lifted between characters. This abrupt change induces a number of harmonics of the fundamental frequency when sending the Morse code dots or dashes and this is heard as a 'click'.

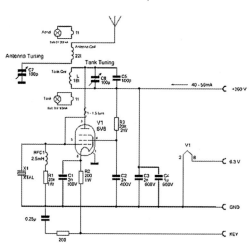

Key click removal circuit (PRJ)

As shown in the circuit diagram, this is easily overcome with the introduction of a resistor and capacitor which have the effect of softening the sharp voltage transition when a character is completed by converting it into a smoothly sloping wave form. Based on the limited operation of the transmitter so far, this modification seems to have been successful as no reports of 'key clicks' have been received.

With the completion of the wiring up, the set can be tested. This involves applying 6.3 volts of alternating current to the heaters of the valves and 360 volts direct current to the anode of the transmitter valve. The receiver runs at a lesser voltage achieved by the insertion of resistors in series with the high voltage supply.

Unlike most transmitters, this radio is not provided with meters but relies on two panel lights which pick up energy from the 'tank' coil of the transmitter. With a quarter-wave length of wire attached to the antenna plug, the set is readily tuned up when the two bulbs shine at maximum brightness. This simple, rapid method of tuning must have been much appreciated by SOE operators in the field.

The loss of power that occurs with energy going to the tuning bulbs rather than into the antenna can be obviated by switching the bulbs out of circuit. Operators were able to achieve a small increase in power output once the set was tuned up simply by unscrewing the two bulbs.

Operation of the Paraset

The principal power supply for the replica is a valve-based laboratory unit. This provides 350 volts DC and 6.3 volts AC for the heaters of the three valves. More recently, a solid state converter that runs from a 6.5 volt Gel-cell battery has been built, and this makes it possible to operate the Paraset as a fully portable unit. With this supply, the standby position on the selector switch becomes important. With the scarcity and high price of metal-cased valves of the type required, heater damage is something to avoid at all costs.

Construction of the Steel Box Replica

The later version of the Paraset (the replica appears on page 186) had the front metal panel and components contained in a metal box. Despite some initial misgivings about working with sheet metal, this was a good deal less of a problem than had been anticipated, particularly with the use of an electric jigsaw.

The simplest choice of metal for the replica would have been sheet aluminium, given its capacity for easy bending and cutting. However, the gauge required to produce a reasonably stiff box would have to be at least 1.5 mm, which would have reduced the sharpness of the bends possible. In the end, 1 mm thick mild steel sheet was used. The result fully justified this decision despite the problems it incurred.

Cutting out sheets (PRJ)

Bending up metalwork (PRJ)

Metal boxes ready for spot welding (PRJ)

Cutting mild steel sheet of this thickness with conventional metal shears is not a practical proposition so a jigsaw equipped with a metal-cutting blade was used to produce the base sheets.

The metal sheet was held down on the work surface with two large G-clamps to negate the vibration generated by the jigsaw. Setting out the straight lines with a metal straight edge and a set square to provide the line of the cut is essential. A metal flat file is required to trim the cut edges of the metal.

The next operation involved bending up the metal using a small pan-brake press which also needs to be held down with the G-clamps. The only tricky part of this operation involved bending up the end plates to the upper and lower parts of the box, and here the slots in the press blade make this possible.

Joining the end plates to the box lid and base was the next step, and soldering was discarded as a means of doing this, as spot welding was used in the original Paraset metal box. A friendly metal fabricator did the necessary work in perhaps ten minutes. The completed box is shown in the accompanying illustration which also indicates the sharpness of the corners and the appearance of the soldered lip to the upper box.

The next step was to mount the piano hinge on the inner surfaces of the

PROJECT 2: Paraset Replicas   185

two box elements. The thickness of the hinge impacts on the available space in the lower box and restricts the space into which the control panel has to be fitted. To save as much space as possible required the piano hinge to be pop-riveted from the inside rather than the reverse (as was the arrangement in the genuine Paraset). The external parts of the 1.4 mm long pop rivets were dressed down to a flat surface to create a reasonably tidy finish. The illustration below shows the fully assembled Paraset case with the piano hinge just visible on the inside surface of the upper part of the box and before the final coating of spray paint.

The last illustration shows the fully assembled Paraset after the application of the dull silver automotive spray paint and set up in the radio room, connected to an external antenna (a G5RV dipole). Sitting on top of the mains-operated power supply is the solid-state battery-driven power pack. Running this latter device from a 6 volt 2.5 amp-hour Gel-cell provides a considerably reduced voltage to the transmitter stage and a corresponding reduction in power output.

When powered from the mains via a laboratory power supply, the Paraset produces a quite satisfactory 4 to 5 watt CW signal. While the Radio Australia service to New Guinea has proved easy to locate at full strength, finding local amateur operators to converse with in CW has proved a good deal more elusive.

The Paraset in the Field

As the various books referred to in undertaking this project made brutally clear, being a radio operator

Finished metal box

in wartime France was a highly dangerous, often deadly, occupation. For that reason alone, the Paraset and its characteristics and operation are of considerable interest and for a modern radio operator seem painfully crude by comparison with today's solid-state miniature devices.

However, compared with the clandestine radios that had preceded it, the Paraset was a great step forward, being comparatively light in weight, capable of being concealed in a conventional suitcase and surprisingly

Second version of the Paraset in metal case (PRJ)

easy to set up and operate. Although the receiver was temperamental and quite difficult to tune accurately, as the regenerative circuit produced considerable interaction with the tuning of the set, the transmitter stage was very effective over the distances required to be crossed in order to get a signal to Great Britain, despite the low power output available.

The Paraset replica represents a fine tribute to those SOE radio operators in war-torn France who risked their lives. It has also found favour with many historically inclined radio amateurs around the world and just Googling the word 'Paraset' will produce an avalanche of sites set up by proud constructors of this interesting radio from the early part of the Second World War.

PROJECT 2: Paraset Replicas

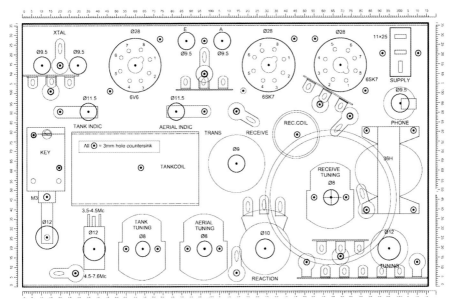

Top panel layout for the Paraset (SM7). Larger version p. 335.

Capacitors

Ref. no.	No. needed	Value in farads	Value in volts	Type
C13	1	10 pF	300	
C9	1	100 pF	100	
C8	1	100 pF	100	
C1	1	100 pF		Variable
C3	1	100 pF		Variable
C10	1	100 pF		Variable
C4	1	100 pF	600	
C2	1	.002 mF	100	
C5	1	.002 mF	400	
C6	1	.002 mF	600	
C7	1	1 mF	600	
C17	1	1 mF	400	
C14	1	.1 mF	400	
C18	1	.1 mF	1000	
C11	1	2 mF	100	Electro
C15	1	2 mF	450	Electro
C12	1	8 mF	450	Electro
C16	1	25 mF	25	Electro

Resistors

Ref. no.	No. needed	Value in ohms	Value in watts	Type
R2	1	200	1	Carbon
R11	1	250	1	Carbon
R5	1	1.5 k	1	Carbon
R7	1	10 k	2	Carbon
R1	1	20 k	1	Carbon
R3	1	20 k	2	Carbon
R12	1	75 k	1	Carbon
R6	1	100 k	1	Carbon
R8	1	100 k	1	Carbon
R10	1	100 k	1	Carbon
R9	2	250 k	1	Carbon
R4	1	1 M	0.25	Carbon

Inductors

No.	Value in henrys	Value in amperes	Type
2	2.5 mH	100 mA	RFC
1	36 H		AFC

Tools and associated items

Small pan-brake press
Metal scribing spike
G-clamps
Large electric drill and bits
Access to spot welding equipment
Pop-riveting tool
Bench vice
Jigsaw with steel cutting blade
Scope Tools high-wattage soldering iron
Metal cutting shears
Flat-faced metal file
Set-square with metric edge ruler

Part 3: 1950–1975

11 After World War Two

Whatever might have been hoped for when the Second World War ended, the reality was that a former ally was to become the new enemy of the West as the Cold War years (1945–1991) unfolded. As Winston Churchill had predicted, the 'iron curtain' that separated the Soviet Union from the West produced a succession of minor and major military campaigns, fuelled by the doctrine of Communism and Russia's drive for world domination.

Atomic Tension

Overhanging all international negotiations in the period post-1945 were the shadows of the atomic bombs dropped on Hiroshima and Nagasaki which had ended the conflict in the East. The apparently unassailable military might of the United States tended to support a dominating and superior attitude in some quarters. While this was perhaps understandable, it was to have potentially disastrous implications for world peace. What was not appreciated at the time was the extent to which the new community of atomic scientists had been infiltrated by Soviet spies, in both the United States and Great Britain. These scientific traitors gained access to the most critical documents, and passed detailed information on the atomic bomb to the Soviet Union. Thus, on 29 August 1949, the Soviet Union was able to trigger the explosion of its copy of the US atomic bomb in the satellite state of Kazakhstan, to the huge consternation of the West.

Further work at the US atomic research centre at Los Alamos in New Mexico led to the creation of the thermonuclear or hydrogen bomb, which was test-fired at Enewetok (now Enewetak) Atoll in the Marshall Islands of the Pacific in 1952. This weapon employed a nuclear-fission trigger to produce the fusion of the hydrogen isotopes deuterium and tritium, with a resultant massive energy release. Again, infiltration by spies and sympathisers enabled the Soviet Union to produce their own version of this terrifying device. A test detonation in 1953 caused massive disquiet in the West and heightened political tensions worldwide.

In addition to the mounting tensions created by the Soviet actions, by late 1949 in China the communist People's Liberation Army under Mao Zedong had overcome the Nationalist forces led by Generalissimo Chiang Kai-shek and set up the People's Republic of China. The Nationalists were forced to retreat across the Straits of Formosa to the island of Taiwan, where they established the Republic of China. This set up the potential for a further major conflict in a region in which the United States and its allies had significant strategic interests. In the midst of this vexed situation, a nation that had traditionally performed a role in separating China from nearby antipathetic nations was to ignite the first serious post-World War Two conflict—the Korean War of 1950–1953.

War in Korea: 1950–1953

Prior to the defeat of Japan, one of the agreements reached by the Allies at a conference at Potsdam in 1945 was that Korea be partitioned along a line that followed the 38th parallel of latitude, a unilateral decision made without Korean consultation or involvement. It was believed that a border between north and south would maintain historical relationships and, most importantly, preserve the 'buffer' nature of Korea in relation to China and its neighbouring states. The boundary was declared by the US State Department-War-Navy Coordinating Committee in Washington, DC during the night of 10 August 1945, four days before the liberation of Korea from Japan. The decision was made without regard for the political pressures that were building up between nationalist forces in the south, headed by Syngman Rhee, and communist elements in the north led by Kim Il-sung.

The competition between the two political systems boiled over into naked aggression when, on 25 June 1950, the Korean People's Army crossed the 38th Parallel and drove south, supported by nearly 300 Soviet-supplied T-34 tanks, rapidly reaching the south-east of the Korean Peninsula.

The South quickly received the support of the United Nations Security Council which on 27 June invited member states to respond to the aggression. The United States was a major participant in the Security Council, and US troops were sent from Japan to assist the Republic of Korea. The headlong advance of the North Korean People's Army was brought to a standstill in August as US reinforcements and weapons boosted the strength of the defenders.

The build-up of troops and munitions enabled the Republic of Korea with its US and United Nations allies to arrest the invasion and in September to initiate a counter-invasion in the north-west at Incheon, the port city of Seoul, capital of South Korea. Here the North Koreans were defeated and Seoul recaptured, and the southernmost RPK forces subsequently driven back behind the 38th Parallel.

Not satisfied that North Korea had been stopped in its ambition to unify Korea as a communist state, on 7 October General Douglas MacArthur, Supreme Commander of the occupying forces in Japan, authorised US troops to cross the 38th Parallel and advance to the Yalu River, the border between North Korea and China, and a highly sensitive location in a geopolitical sense. MacArthur apparently believed that the Chinese government would not react. This was a grave misjudgement. The Chinese reaction was both violent and a major surprise to the southern allies. On 19 October, the People's Volunteer Army (PVA) of China crossed the Yalu and the next phase of the Korean War was launched.

Communist China was able to put a huge infantry army into the north of Korea, but it was characterised by relatively small logistic support and lack of modern artillery, tanks and telecommunications. US forces were smaller in number, but supplied with apparently inexhaustible quantities of tanks, trucks and supplies with which to support its troops in the field. Thus while the PVA had to rely on the endurance and even fanaticism of its infantry to take the fight to the ROK and US troops, the Allies tended to place rather too much confidence in armour, artillery and logistic support, and rather too little in developing the fighting spirit of the men on the ground. The expression 'bugging off', meaning a hasty retreat having abandoned arms and ammunition, began to appear in reports relating to defensive actions in which Chinese forces had obtained the upper hand.

Following the intervention of the PVA and its initial stunning success in driving back Allied forces, an unexpected pause occurred as the Chinese withdrew into the mountains and waited to see how the Allies would react. The outcome was a further assault by ROK troops along the eastern and western flanks of North Korea, but the PVA overcame both forces and saw them retreat southwards once again. This process continued until the PVA recaptured Seoul in early January 1951.

At this stage, General Matthew Ridgway was appointed to lead the US Army in Korea as a replacement for his predecessor, General Walker, who had been killed in a motor vehicle collision. This change in the command structure saw the stemming of the southwards retreat of the US forces and the start of a process of attrition in which the infantry forces of the North and China were finally brought to a halt by the sheer weight of US-supplied armaments and men. During this period MacArthur, who had seemed intent on generating a full-blown war with Communist China, was relieved of his position as Supreme Commander in Korea. His threats to employ nuclear weapons and his advance to the Yalu River, despite presidential orders that such a step should only be taken in the face of offensive action by China, had convinced President Truman that MacArthur's removal was essential if peace were to be achieved. Ridgway was now responsible for the conduct of the war and oversaw the development of a stalemated situation horribly reminiscent of the trench warfare of the First World War.

This stage of the Korean War commenced in May 1951 and continued until the signing of an armistice in 1953. During this period, thousands of troops from both North and South were to die as the negotiations for the cessation of hostilities ground forward at a snail's pace and frequently stalled.

Finally, in July 1953, an armistice agreement was reached and the guns ranged on either side of a line that followed approximately the 38th parallel of latitude at last fell silent. This was to prove a somewhat problematic cessation of hostilities, however, as a peace treaty had not been negotiated.

Since 1953, the two Koreas have existed in a very uneasy relationship on either side of the Demilitarised Zone established as a result of the armistice agreement. There have been a number of incursions from North Korea and that state has also gained access to nuclear weapons, adding significant tensions to the relationship. As well, while South Korea has gone ahead in terms of industrial and social prosperity, the North has existed in a state of grinding poverty and political oppression. The replacement of the first North Korean leader, Kim Il-sung, by his son Kim Jong-il, and in turn his grandson Kim Jong-un, has not significantly changed the political situation or relieved the tensions that existed in 1953.

Australia and the Korean War

While Australia's contribution to the allied effort in Korea was comparatively small as a percentage of the nation's population, the 17 000 servicemen who served there between 1950 and the armistice in 1953 represented a significant component of the Security Council-sponsored Allied forces. The Australian Army, Navy and Air Force sustained 1500 casualties and 339 combat deaths in total.

When the hostilities in Korea broke out, Australia's main complement of troops was based in Japan as part of the international occupation force. Immediate support for the UN-backed counter-invasion force was provided in the form of No. 77 RAAF Squadron, flying P-51D Mustangs. These were propeller-engine fighters of the Second World War era that had been upgraded to respond to conditions that pertained in 1950.

Unfortunately for the Australians, it was not long before Chinese-flown MiG-15 Soviet jet fighters took to the skies over Korea. This precipitated No. 77 Squadron's rapid conversion to British-designed Gloster Meteor jet fighters. These proved to be a good deal less agile in combat than the MiG-15 and No. 77 Squadron was soon relegated to ground attack and support roles as US Sabre jet fighters took on the aerial combat role. However, with access to radio communications, the Meteors were able to supply overwhelming support during infantry engagements, where the Soviet-built T-34 tanks, in particular, proved vulnerable to air-launched armour-piercing rockets.

The Australian Army soon followed the RAAF to Korea, in September 1950, and the 3rd Battalion, Royal Australian Regiment (3RAR) entered battle as a component of the 27th British Commonwealth Brigade. Over the next two years, 3RAR was intensively engaged with the UN-sponsored force, which included as its principal element the US Army. This relationship underlay the awarding of a US Presidential Unit Citation to 3RAR following a particularly gallant engagement with the Chinese Communist Army during the Battle of Kapyong, which led to a major local victory. This occurred on 23 April 1951, when a determined assault from the north was stalled as a result of dogged and desperate opposition by the Australian forces. Coincidentally, this action occurred just two weeks after President Truman's dismissal of General MacArthur and his replacement by General Ridgway.

Later in 1951, 3RAR again distinguished itself, during the Battle of Maryang San in which a Chinese Communist Army position at a bend in the Imjin River was subject to attack. During this action, 3RAR was able to dislodge a numerically superior force from a strongly defended position, in the process suffering 20 combat deaths and nearly 90 casualties.

During the ensuing two years until the armistice was arranged, in common with other members of the UN forces, 3RAR was compelled to participate in the static trench warfare so reminiscent of the First World War.

While the Korean War has been characterised as the 'Forgotten War', due to its relative brevity and the somewhat uncertain political outcome that still lingers over the Korean peninsula some sixty years later, it was to have some very significant implications for Australian international relationships. One result of the Australian involvement in Korea was the solidification of the ANZUS Treaty of 1951 which established a special new relationship between Australia, New Zealand and the USA, which remains in force.

Radio Communications in Korea

A particular feature of the Korean War was the disparity of equipment and military apparatus between the Republic of Korea Army with US Army support, and the North Korean Army operating in conjunction with the Chinese People's Volunteer Army. No more extreme was this contrast than in the availability and use of radio for communications, particularly in the field and in support of infantry forces.

Where the US Army and UN-sponsored forces had access to HF and VHF radio transceivers for their front-line troops, the Chinese Army had to resort to various acoustic signalling systems, such as gongs and whistles, to convey orders to the front line. When first encountered by the defending forces, these tactics caused considerable anxiety, but the troops quickly became used to the acoustic onslaughts. With the defenders' ability to use radio to call up artillery or airborne support, the comparison between technologies could not have been more stark.

By the time of the Korean War, the clear advantage of VHF and FM systems over earlier HF communications equipment for short-range communications was well established. The British Army had moved on to a new range of radios referred to as the Larkspur range, but these had not been distributed to British Commonwealth forces by the time the Korean War began. The Australian Army in Korea operated in the main as a component of the much larger British-led force and the radios in use were those supplied to Commonwealth members at large and Great Britain in particular.

During the early part of the Korean War, the British Army still relied on HF-based radios such as WS No. 19 and the later, closely related WS No. 62, while the US designed Handie-Talkie, the BC-611, was still available to the Commonwealth forces. Confined to a single HF channel, this was a very useful short-range transceiver despite this limitation (see Chapter 8). US forces on the other hand were supplied with the BC-1000 VHF FM radio (also discussed in Chapter 8).

Wireless Set No. 19 (Aust.) (RAS)

The use of HF by Commonwealth forces and VHF FM by the US Army resulted in serious operational incompatibility, a major difficulty when US tanks were used in concert with British, Australian and Canadian infantry forces. Absence of a direct radio link was an often fatal problem in the face of a common Chinese method of attacking US tanks—which involved clambering onto them and pushing explosive packs through open vents. Because soliciting supporting infantry fire by radio contact was not an option, the US tank teams had to rely on machine-gun and cannon fire from neighbouring tanks in their formation to clear off the attackers. If a common communications frequency had been available, a defensive response could have been accomplished a good deal more efficiently by infantry small-arms fire.

Lack of a common calling frequency was overcome later in the war when new VHF transceivers were brought into use by Commonwealth forces, in particular, the British VHF backpack transceiver WS No. 31. Able to communicate on frequencies between 40 and 48 megahertz, the WS No. 31 was able to talk to US troops who were supplied with the backpack

BC-1000 from which it was derived. This VHF transceiver covered the same range of frequencies as the WS No. 31.

Other VHF pack sets progressively introduced for use by Commonwealth and Australian forces included the British-designed WS No. 88 and the Canadian C/PRC-26.

The characteristics and limitations of RF propagation at HF and VHF became very apparent with the introduction of these new radios. The hilly, rugged Korean terrain tended to reduce the usefulness of VHF radios where 'line of sight' communication was not possible, and despite the inefficiency of whip antennas at HF frequencies, HF radio propagation tended to be less adversely affected. This was particularly the case where HF groundwave radiation was significantly less affected by topographic changes than at VHF. As also became apparent, in areas of dense vegetation and jungle canopy, HF radio could be very useful, particularly where more distant operations were concerned.

Where more efficient wire aerials could be used at HF, it was possible to employ high-angle ionosphere-reflected radiation. This mode of operation could make possible effective communications over a substantial radius of reception, beyond the penetration of local ground-wave radiation from low-

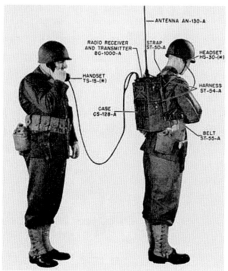

Wireless Set No. 62 (RAS)

BC-1000 VHF FM (USM)

powered portable transceivers. This was particularly useful where operations were undertaken under a heavy jungle canopy and a clearing was available in which to erect a dipole set up for the frequency in use.

Given the vagaries of RF-based communications at VHF and HF, the extent to which conventional wired telephone was used in Korea is perhaps unsurprising, despite its vulnerability to motorised traffic and the physical problem of laying out the cables. Army signallers were just as busy laying twin telephone cable in Korea as they had been in the Second World War, although radio was frequently used in parallel or as a substitute when the lines were severed.

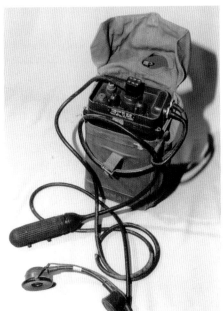

WS No. 88 VHF (PRJ)   C/PRC-26 VHF (PRJ)

The Malayan Emergency: 1948–1960

While French efforts to retain their colonial interests in French Indo-China, now Vietnam, had been faced with determined resistance led by communist forces after the end of the Second World War, in Malaya it was a generally different situation. Although nationalist feeling supported a drive for independence from Great Britain, communist incursions generally involved discontented ethnic Chinese Malayans, who tended to control business activity but represented a minority. The dominant ethnic group was Malay, with the Chinese the second largest group and the Indian component of the population a distant third.

During the war against Japan, British forces had worked in Malaya in concert with the Malayan People's Anti-Japanese Army, a communist resistance force. It was this group that was to turn against Great Britain and its commercial interests at the ending of the war. The first serious expression of this opposition was the assassination of European rubber plantation managers in the northern state of Perak in 1948. Thus the so-called Malayan Emergency could be seen to have commenced before the war in Korea broke out. The main reason for referring to the conflict as an 'emergency' rather than a 'war' related to commercial insurance policies that would have been invalidated if war were declared.

With the benefit of hindsight, and despite the semantic juggling, there can be little doubt that the guerrilla campaign that was waged by Chin Peng, the leader of the Communist forces in Malaya, did indeed constitute a war. This war continued until 1960, well after the creation of the independent Republic of Malaysia in 1957, and had earlier led to the departure of British and Commonwealth forces from the peninsula.

Australian troops were brought to Malaya in 1955 where they worked with other elements of Commonwealth military forces, being referred to as the British Commonwealth Far East Strategic Reserve. This was a very difficult campaign conducted in largely jungle terrain which demanded the application of specialist tactics and the Australian Army was required to re-learn techniques of jungle warfare that had been taught at the Canungra Military Area in south-east Queensland during the Second World War. This facility was re-opened in 1954.

In October 1951, the British High Commissioner, Sir Henry Gurney, was ambushed and killed by the Communist forces. He was replaced by Lieutenant-General Sir Gerald Templer. This change was to have substantial longer term benefits because Templer brought to Malaya an approach which included what would now be called social engineering. This was designed to win the 'hearts and minds' of the Malay population and as a particularly effective step involved the protection of village settlements from communist insurgents. This was achieved through a process of village resettlement in which the villagers were moved away from their traditional kampongs to new secure compounds where ingress and egress could be controlled. Resettlement was originally opposed, but as the benefits of stability and increasing affluence became apparent the relocation process appears to have been generally accepted. The military response to the communist incursions was to institute an active process of patrols and ambush. These patrols, undertaken at considerable distances from the military base of operations and conducted under the thick canopy of the jungle, had significant repercussions for the forms of radio communication that could be employed.

Jungle Communications

During the Korean War, the US Army had taken delivery of newer, more compact VHF transceivers, the AN/PRC 8, 9 and 10, and after the war the Australian Army had also been equipped with this apparatus. Significantly smaller than the BC-1000, these three radios covered a range of frequencies between 20 and 54.9 megahertz. The PRC 8 ran from 20 to 27.9 megahertz, the PRC 9 from 27 to 38.9 megahertz, and the PRC 10 from 38 to 54.9 megahertz. A small frequency overlap was available between the adjoining pairs.

Superficially the three radios were identical, with frequency selection made via a tuned variable frequency oscillator (VFO) and a battery pack that supplied both low and high voltages to drive the miniature valve complement. The battery was located in a compartment below the transceiver casing. These radios were designed to operate with flexible whip antennas and produced around 1 watt of RF power which allowed a range of communications of about 5 kilometres.

However, as was quickly discovered, VHF FM radios work best where close to a 'line of sight' relationship between apparatus is possible. In heavily vegetated country and jungle areas the radio frequency energy is rapidly attenuated and contact is often prevented. The output power of field radios operating in the VHF part of the RF spectrum was quite small at the time and, even under the best conditions, they were essentially quite short-range devices.

AN/PRC 10 VHF FM (PRJ)

In the face of such problems, HF radio once again became relevant— but operated in a different physical configuration to the backpack with whip antenna used during the Second World War with groundwave propagation.

In the jungle environment it was necessary to rely on skywave, in which RF energy propagated at a high angle was

A510 in carrying case with hand generator (PRJ)

Wireless Set A510 with whip antenna coupler (PRJ)

reflected from the ionosphere, and for which wire aerials were used. Clearings and open areas were essential launching sites for long-distance communications.

Obviously a lightweight HF radio was needed, and the Australian Army specified new miniature valve-based equipment that was designed and supplied by AWA. This was the A510, a new compact radio split into two elements, the transmitter and a separate receiver. Unlike a backpack radio, the A510 was contained in two belt-mounted pouches which, among other things, made the signaller a much less obvious target for enemy sniper fire. Dry batteries were contained in detachable compartments below the two parts of the radio.

Supplied in a wooden box which contained almost all the accessories required to set up a portable HF station, this radio was found to be reliable and with it, encoded Morse code could be sent by means of CW. The A510 was quite low-powered but nevertheless very successful in maintaining communications over 30 to 40 kilometres using wire dipole antennas.

British Military Radio: Larkspur

In the late 1950s, the influence of British Army signals equipment remained strong in the Australian Corps of Signals, although knowledge of US radio design produced a fundamental change in Australian thinking in the run-up to a new war in Asia: Vietnam.

The venerable WS No. 19 and the more recent WS No. 62 were then being relegated to training purposes and use by the Citizen Military Forces. Their replacements, from the new British Larkspur range, were designed to overcome a number of basic problems in the older equipment. In particular, the problem of moisture ingress was tackled very effectively.

As the illustration overleaf of a Larkspur VHF radio demonstrates, Wireless Set C42, its power supply and antenna tuning unit, were contained in diecast aluminium cases, with substantial ribbing for added strength. Less obvious is the rubber gasket between the front panel, carrying all the controls and metering, and the main body of the radio. This radio was capable of short-term water immersion. For an equipment restorer, a very pleasant surprise results from opening one up for service many years after its introduction, for the hermetic sealing has the effect of preserving the internal components in pristine, virtually original condition. This can make a startling contrast with the external condition, which frequently exhibits the results of hard active service.

Wireless Set C42 VHF FM: Larkspur equipment (PRJ)

The WS C42 transceiver had a frequency range of 36 to 60 megahertz, and employed double and triple conversion in a superhet receiver. The transceiver had continuous tuning over the band of frequencies covered, with the channels marked on a moving film that showed in the tuning dial opening. A crystal calibration unit allowed the frequency required to be accurately set.

As vehicle-mounted equipment with access to the battery, the radio's power supply was designed to run on 12 or 24 volts and generate between 15 and 20 watts of RF power. Switched to low output power, it produced 0.5 watts.

In addition to the VHF radios in the Larkspur range, a new HF transmitter-receiver pair was introduced with the designation C11-R210. Contained in cases very similar to the C42 VHF transceiver, this radio had a frequency coverage of 2 to 16 megahertz with three switched ranges. The C11-R210 was able to produced amplitude modulated (AM) speech and CW-based Morse code. Being vehicle mounted with access to lead acid battery power, the transmitter had a power output of 45 watts high power or 10 watts low

C11-R210 HF transmitter and receiver: Larkspur (PRJ)

power and could provide AM, CW or phase-shift keying for teletype.

12 Computers, *Sputnik* and ARPANET

Before the start of the Second World War, a highly secret organisation was set up in Great Britain to decode German radio messages received at a series of listening stations in England. Known then as the Government Code and Cypher School, and after 1946 as the Government Communications Headquarters (GCHQ), this now famous organisation was established to the north of London at a mansion and estate known as Bletchley Park.

Electronic Computing

One of the key war-winning technologies developed at Bletchley Park involved new computational machines that assisted in the complex and repetitive activity needed to break the German codes. Most secret of these was a machine to assist in deciphering encoded teletype signals, the method used by the German High Command to distribute top-secret instructions to its distant armies and assumed to be unbreakable. The new British computing machine, known as Colossus, employed an array of over 2000 valves and was the forebear of a whole host of post-war computers. Due to the secrecy surrounding Colossus, for more than thirty years after 1945 it was not generally known that this first valve-based machine had initiated the new technological science of computation. The post-war computer industry arose with no overt recognition of the developments at Bletchley Park.

German assumptions about the security of the coded output of their two encoding teletype machines, the Geheimschreiber ('secret writer') and the Lorenz encoding machine, were far from justified. With Colossus operating in conjunction with another machine known as Tunny, the British were able to break the incredibly complex codes and some of the most important strategic decisions that Hitler was to take during the Second World War were read at Bletchley Park.

Creating a new method of computation for codebreaking led to the first generation of post-war electronic computing machines and in turn to a series of technological developments that have fed into military telecommunications up to the present. In particular, the expanding use of solid-state devices and integrated circuits revolutionised battlefield communications. More recently, the introduction of data communications networks for enhanced command and control have produced capabilities quite unimagined even in 1970. These issues are discussed in greater detail in Chapter 14.

Cold War Sparring

During the latter part of World War Two, while the United States and Great Britain were intensively engaged in developing the atomic bomb, German scientists were intent on producing 'vengeance weapons' or 'V' bombs, the deadly V-1 'Doodlebug' and the high-altitude rocket-propelled V-2. When the V-2 was fired into the fringes of space on 8 September 1944 and followed a trajectory that saw it land in the London suburb of Chiswick, killing a number of people, the stage was set for the post-war development of a hideous new weapon: a ballistic missile carrying a nuclear warhead. Germany's V-2 opened up the possibility not only of space flight but also of nuclear Armageddon. This potential was to underlie much of the international tension that fuelled the Cold War.

As the Cold War unfolded, the arms race it encompassed revealed a terrifying portent for the future of Europe and North America. At the same time, tensions between the West and the Soviet and Chinese communist blocs found expression in small-scale but entirely 'hot' wars in Southeast Asia. The Korean War came very close to initiating a new world war, a war employing nuclear weapons. The recall of General MacArthur averted that disaster, but the fear that nuclear warfare could erupt elsewhere lingered, fuelled by the development of high-altitude rockets for space exploration. These events were imagined as a prelude to new means of delivering nuclear arms.

Nuclear Developments and *Sputnik*

Only four years after the detonation of the US atomic bombs at Hiroshima and Nagasaki, on 29 August 1949 the Soviet Union exploded its first nuclear device in Kazakhstan. Following this triumph of stolen technology, on 12 August 1953, just after the end of the Korean War, the Soviets detonated a thermonuclear device there, producing paroxysms of concern throughout the West. Heightened suspicions of the intentions of the Soviet Union and fears of the advance of communism created a manic environment in the United States in which Senator Joseph McCarthy was able to hold hate-filled public enquiries into the political beliefs and loyalty of US citizens.

Only a few years later, the Soviet Union established its priority in the developing 'space race' by launching the first Earth-orbiting satellite. In the United States this was seen as yet another step towards communist supremacy and a new world war, and contributed to a steady progression to war in Southeast Asia.

In 1954, when France was forced to surrender to Communist forces at Dien Bien Phu in Indochina, the expanded involvement of the United States in the region was close to a foregone conclusion. Australia would be drawn into the Vietnam War because of both the special relationship with the United States confirmed with the signing of the ANZUS Treaty in 1951, and its membership of the South East Asia Treaty Organization (SEATO), established in Manila in 1954.

On 4 October 1957, the United States awakened to a new phenomenon, a 'bleeping' radio signal from the edges of space. This was the voice of *Sputnik 1*, the first man-made, low Earth-orbiting satellite, which transmitted a data stream on two frequencies, 20.005 and 40.002 megahertz. These Soviet signals could be heard by US government agencies and by thousands of radio amateurs around the world who recorded and rebroadcast the sound.

US fears of communism's insidious intent of world domination would find no better focus than this small space vehicle with its insistent radio signals. The response was a demand for extreme action and the development of means to combat the perceived potential for space-based nuclear attack. From this came a series of developments that would culminate in one of the most important elements of modern life, the Internet.

The precursor of many of these developments was the creation in the United States, several years pre-*Sputnik*, of the Air Defense Systems Engineering Committee under the chairmanship of Professor GE Valley (often known as the Valley Committee). This committee was to advise on air defence systems and with

Sputnik 1: the first satellite (PRJ)

proposing the best means to respond to the perceived threat from the Soviet Union and its demonstrated capacity to wage nuclear warfare. The outcome of the Committee's deliberations in 1950, a proposal for the automated air defence of North America known as the Semi-Automatic Ground Environment or SAGE, led to the creation of some 23 distributed communications-linked computing centres which progressively became operational from 1958. The computers were built by IBM, and this defence contract was as much as anything else responsible for establishing IBM as pre-eminent in the burgeoning US electronics industry.

ARPANET

Another reflection of the post-*Sputnik* panic was the creation in 1958 of a new entity in the Department of Defense in Washington, on the initiative of President Eisenhower. This organisation was known as the Advanced Research Projects Agency (ARPA), and within it was a section directed at information processing techniques, the Information Processing Techniques Office (IPTO). Initially IPTO was led by the psychoacoustician JCR Licklider, who was recruited from the architectural acoustic consulting firm Bolt Beranek & Newman. Later, in 1966, R (Bob) Taylor took the helm and prepared a design for a distributed network of communications linkages across the United States to respond to the perceived vulnerability of the existing elongated linkages provided by hard-wired, decentralised transcontinental telephone lines.

DECENTRALISED NETWORK

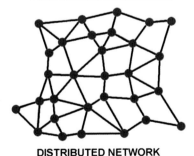
DISTRIBUTED NETWORK

Decentralised and distributed networks (PRJ)

In concert with conceptual work on the distribution of digital information as data 'packages' or 'packets', carried out at the RAND Corporation by Paul Baran, was the notion of sending such information via a distributed network of wired links in which a multiplicity of alternative routes was available between origin and destination. Instead of the data being conveyed as a continuous stream of information, data packets could travel as single entities by various routes, to be reassembled into their correct order at the receiving end of the communications path.

This approach was coincidentally very similar to work being carried out in Great Britain by the National Research Laboratory at Teddington under the supervision of Donald Davies, in a scheme developed in concert with British Telecom, which was very keen to see the national system of hard-wired telephone cables used more effectively via digital techniques. From Davies' work came the highly significant concept of 'packet switching'—something quite distinct from the operation of computers—that made use of the network to convey data that they had created to other computers.

Despite AT & T's initial lack of enthusiasm for the project in the United States, it was soon appreciated that to make a system involving the transfer of digital information as data packages, a special-purpose computer would be necessary—and one that was independent of the machines it was designed to provide linkages to and from. From this realisation came the development of the interface message processor (IMP), built in response to a specification sent to a large field of manufacturers in 1968. The military contract went to Bolt Beranek & Newman, a familiar name, who had made use of the recently developed Honeywell minicomputer, the DDP 516, for which processing software was produced using Assembler language. These new machines became the nodes in the new ARPANET data communications network, which grew from an initial group of four connected university-based IMPs in 1969 to twenty-four in 1972 and continued to expand exponentially for many years thereafter, until it was ultimately superseded by a far faster network, the Internet.

Part of the success of the ARPANET was the development of the data transfer protocols which today underpin the Internet. The initial network protocol of 1970 was succeeded in 1972 by the file control protocol (FTP), which can still be encountered from time to time. This protocol allowed disparate computers involving a variety of hardware configurations and manufacturers to organise data for communications to other computers on a consistent and unambiguous basis; from it, very quickly, came an avalanche of electronic mail. Based on programs developed by Ray Tomlinson, the familiar @ symbol was introduced as a separator in the email address information.

It was soon apparent that connection to international systems, particularly in Great Britain and Europe, was necessary and from this came the need to expand the addressing protocols to accommodate satellite linkages and, later, undersea cables. In order to develop the necessary

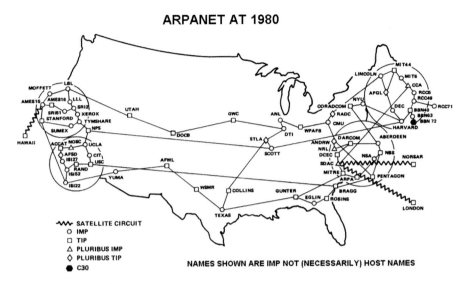

ARPANET at 1980 (INT)

software, Vinton Cerf of Stanford University in California developed what were to become the key protocols in international networking, the transmission control protocol (TCP) and the internet protocol (IP), which combined are the familiar TCP/IP used to identify any Internet unique resource locator (URL). This system became the basis of ARPANET in 1983, and the Internet later adopted the same system.

Military Networks: Command, Control and Communications

By 1945, military notions of what is now referred to as 'command and control' methodology were well established. The need for integration of communications, particularly between forces in the air and forces on the ground, was appreciated. In a hierarchical military organisation, which in management jargon would be described as pyramid shaped, at the apex of the pyramid were located the commanding generals and at its base were the platoon commanders responsible for converting orders into action.

Orders reflecting strategic ambitions were issued at the top of the pyramid and percolated down to the fighting forces at the platoon level, where they led to operational planning and detailed tactical orders. Planning and the resultant orders had to take into account the constraints of local conditions and the actions of opposing forces. The expectation was that orders would be followed without excessive deviation from the strategic objectives determined at the apex of command and without questioning the strategic intent. Where air support was called up in support of ground forces, orders were passed up and down the respective functional groupings, which sometimes caused delays and misdirection. Later, attaching airborne combat-support aircraft to particular elements of the infantry worked successfully, despite the logistic difficulties inherent in decentralised forces operating away from a central command and servicing facility. By the time of the Korean War, these two diametrically different approaches were causing significant administrative difficulties that led to a long process of structural re-evaluation and adjustment which in more recent times also has had to respond to unconventional military and paramilitary threats.

The wars in Southeast Asia required re-examination of some the fundamental assumptions and arrangements developed prior to 1945. The wholly different terrain and weather conditions to those of Europe, coupled with confronting opponents whose tactical approach was often unconventional, generated significant initial problems. No more extreme were these new conditions than in the coordination of air- and ground-based combat in Korea.

Thus in the new battlefields of Korea and in the jungles of Malaya and Indonesia, a significant degree of flexibility and autonomy was required in the infantry context. The notion that platoons in massed arrays were the fighting front of a pyramidal-shaped command structure began to erode. More and more, the detailed development of military actions had to be left to the discretion of field commanders, and high-grade communications from battlefront to headquarters became an essential component of command and control.

Painful experience was to confirm that the 'fog of war', a term introduced in 1857 by Prussian military analyst Carl von Clausewitz, tended to make the best-laid plans of the most exalted generals go awry, particularly in the jungle situation, and in the face of the unpredictable actions of insurgents and guerrilla forces. The old notion of the 'proper' way to fight a battle became largely irrelevant with the rise of asymmetric guerrilla warfare.

In Korea, where regular United Nations and US forces came to face an unconventional military opponent for the first time, Second World War methods of command and control were a good deal less successful than might have been expected. Most particularly this was true of the disposition and support of aircraft in relation to ground attack. The end of the conflict in 1953 saw a very thorough re-examination of military doctrine and methods of command, control and communications.

Closely associated with the acknowledgement of new issues in command and control was the realisation that in the field of communications, serious issues of compatibility and interoperability had to be confronted. The need for ground troops to access airborne resources and artillery demanded communications apparatus capable of providing the linkages for successful contact. The most obvious points of concern were the range of frequencies covered and whether amplitude or frequency modulation should be employed. At a later stage, the compatibility of digital methods of communication had to be ensured.

From the post-Korean War reassessment came new approaches that led to something akin to the Baran form of dispersed network in which all elements of a military command were linked by instantaneous high-grade communications supplied via a network. For this a totally new technology was required, and between 1950 and 1970 computerisation began its extraordinary growth. With this expansion came the development of computer-based communications and data transfer (discussed in Chapter 14).

Concurrent with the development of ARPANET for military purposes, in both universities and the commercial market first-generation computers were being linked together in local area networks (LANs). The military later embraced this technology as essential to the complex demands of future combat situations, whether on land or sea or in the air. When the prospect of needing to respond to threats from the edges of space arose, with very limited time to react, the new technology would provide the only effective means of response. In the launch and flight of *Sputnik 1* can be seen the first traumatic step towards the networked age in which we now live with the Internet as its most obvious expression.

As had happened in the past, with the threat of possible major conflict technology would receive a boost. New developments in telecommunications would have huge repercussions in both the military and civilian spheres in the years to come.

13 The Vietnam War: 1959–1975

As the war in Korea ground to its bitter end, another major conflict was in prospect as France attempted to reinstate its position in the colony of French Indochina.

From the eighteenth century and sometimes earlier, a number of European powers had established colonial administrations in Southeast Asia. During the 1930s the military expansion of Japan and its efforts to establish the Greater East Asia Co-Prosperity Sphere were responsible for breaking down much of the European colonial influence in the region.

At the end of the Second World War, the defeat of Japan acted as a signal to various nationalist groups to become vigorously active in efforts to ensure that the colonial powers did not regain their possessions, and they expanded into the vacuum left by the removal of the Japanese aggressors.

Vietnam was one such example, with the added concern for the West that here nationalistic fervour was associated with a developing communist power base and a Communist Party leader who in the pre-war period had come to represent pent-up anti-colonial frustrations. The leader was the charismatic Ho Chi Minh and the colonial power he wished to see displaced was France.

With Ho Chi Minh's declaration in Hanoi in September 1945 that Vietnam was free of French rule, the stage was set for a 30-year struggle. Initially the battle was with the French who strenuously sought to regain the position that they had enjoyed before the Japanese invasion. However, by 1945 much of the strength of the French empire in Southeast Asia had been eroded and its capacity to resist the pressures for independence had been greatly diminished.

The military campaign that followed the declaration of independence came to a head in May 1954. Despite considerable logistical support from the United States, the French Expeditionary Force was besieged for two months at its major military base at Dien Bien Phu in the northwest of the country and forced to surrender to the communist Viet Minh forces. Shortly afterwards the 1954 Geneva Accords were signed, under which France agreed to withdraw its forces from all its colonies in French Indochina—Vietnam, Laos and Cambodia—while stipulating that Vietnam be temporarily divided at the 17th Parallel. In a situation reminiscent of the division of Korea, control of the north was given to the Viet Minh as the Democratic Republic of Vietnam (DRV) under Ho Chi Minh, with the south becoming the State of Vietnam under the former emperor Bao Dai, preventing Ho Chi Minh from gaining control of the entire country. The United States continued to provide aid to the government of Bao Dai.

Bao Dai was shortly replaced as head of state by Prime Minister Ngo Dinh Diem, whose government ruled the increasingly troubled south (now officially the Republic of Vietnam) for the next nine years until he was assassinated in a coup in November 1963. The Ngo Dinh Diem government represented a Roman Catholic minority in a predominantly Buddhist state, and part of the escalation of tensions can be directly related to its persecution of the Buddhists. This religious tension finally came to a head when a number of Buddhist monks self-immolated as a gesture of protest, thereby encouraging their already antagonistic fellow clerics.

The demise of Ngo Dinh Diem and his replacement by the first in a succession of army generals was hastened by increasing incursions from the north along the notorious Ho Chi Minh Trail. Human portation of military arms and supplies along the route supported the expanded activities of the communists in the south. This new presence, officially the People's Liberation Armed Forces (PLAF), opposed the Army of the Republic of Vietnam (ARVN). Its early successes caused major concern in the United States and led to the provision of military support to the ARVN. By 1962, the United States had expanded its forces in the south to some 11 000 'military advisers'. The Viet Minh who had ousted the French at Dien Bien Phu became the Viet Cong, a force that would be long remembered by servicemen of both the United States and Australia as a brutal opponent, although acknowledged as apparently fearless in the face of Western firepower.

President John F Kennedy appears to have recognised that the war was only likely to be won by the South Vietnamese themselves, not independently by the United States through its armed presence. Had his view prevailed and guided the US approach, there might have been a very different outcome. But Kennedy was assassinated in 1963 and the war in Vietnam ground on remorselessly until 1975, with the loss of thousands of US and Vietnamese lives, and substantial numbers of Australian and New Zealand service personnel.

An incident in the Gulf of Tonkin, in which in which it was said that the USS *Maddox* was attacked by three North Vietnamese vessels, and a number of lives were lost in the exchange of fire that followed, became the rationale for the United States to expand the numbers of ground troops sent to support the South Vietnamese government. Ultimately more than 500 000 infantry and support personnel were involved.

In the US commitment to South Vietnam, the driving concern was to prevent the consolidation of North and South Vietnam as a communist state. This was perceived by strident anti-communist forces in the United States and elsewhere as the first step in a process referred to as the 'domino effect' in which the whole of Southeast Asia was seen as likely to follow the Vietnamese movement towards communist ideology, with the initial collapse precipitating the collapse of the next nation in the chain.

Given Australia's proximity to the Southeast Asian nations seen as likely subjects of communist expansion, it is no surprise that the US position was at first generally supported by the public and politicians. The commitment to Vietnam in 1962 of a limited number of military advisers by the government led by Robert Menzies was relatively uncontroversial. But in 1965, when the first contingent of infantry was sent to Vietnam, the response from the Opposition was both pessimistic and hostile. Over the next seven years, public antagonism towards this Asian war was to increase to the point that withdrawal of the Australian contingent became inevitable.

As the United States and its allies were to discover, this was not to be a conventional war of the sort so recently fought in Europe, but one in which the enemy was an elusive target and in which there was rarely a front line. Not only that, the enemy included many apparent civilians and effectively enveloped the Allies, employing guerrilla tactics that US forces seemed singularly ill-prepared to counter. The US response was conventional, involving the use of massive firepower on the ground and in the air—which tended to engender hostility among the civilians it was intended to protect.

What ensured the North Vietnamese victory and the withdrawal of the US Army and its allies was the failure of political support in the United States and elsewhere because of the overwhelming perception that the war had been lost. An armistice agreement, the Paris Peace Accords on 'Ending the War and Restoring Peace in Vietnam', was signed on 27 January 1973. Ironically, the immense drain of manpower and resources in the south that resulted from the continued communist attacks did not constitute a true victory for the North. Only later, when South Vietnam lost its capacity to defend itself as a democratic power in a political sense, was the North able to launch a victorious final assault, contrary to the terms of the armistice agreement. The ARVN resistance disintegrated, and on 30 April 1975, troops of the Vietnamese People's Army entered the city of Saigon. South Vietnam became a component of the wholly communist united Vietnam.

Australia in the Vietnam War

Although a very small element of the Allied build-up in Vietnam, Australia provided a contingent of troops, initially in the form of thirty military advisers led by Colonel FP (Ted) Serong which arrived in Saigon in July 1962 and was referred to as the Australian Army Training Team of Vietnam (AATTV). In August 1964 the RAAF sent a flight of Caribou transports to Vung Tau. In June 1965 a company of the 1st Battalion of the Royal Australian Regiment (1RAR) was attached to the US 173rd Airborne Brigade, located some 25 kilometres from Saigon in Bien Hoa Province.

Because of fundamentally different approaches to the task of holding territory, the Australian infantry was later treated as a task force and given a specific area of responsibility. In April 1966 the 1st Australian Task Force (1ATF), consisting of the 5th and 6th Battalions of 1RAR, was sent to Phuoc Tuy province, where a base was established at Nui Dat. In August this force was the subject of a major assault and the battle that erupted in a rubber plantation close to the village of Long Tan, in the vicinity of the base at Nui Dat, was to enter the annals of Australian military valour. While patrolling to the east of the base, the leading platoon of D Company of 6 Battalion RAR came into contact with a substantial force of North Vietnamese Army troops and Viet Cong, estimated at perhaps 2000 men. The battle that followed saw the vastly outnumbered Australians beat off the encircling attack. The arrival of a number of armoured fighting vehicles to assist the embattled patrol led to a clear victory for the Australian forces and established them as the dominant force in Phuoc Tuy province, a situation that was maintained until withdrawal six years later.

In addition to the remarkable heroism of the defending Australian forces in this action, what stands out is the importance of the part played by military radio technology. Radio was the means by which the forward observation officer (FOO) was able to direct the artillery defensive fire when successive waves of communist forces stormed company headquarters. The successful maintenance of a wireless link between the FOO and the New Zealand artillery battery at the base compound ensured that defensive fire was able to be brought down within 50 metres of the front line with quite devastating impact on the enemy.

1RAR was to remain in Vietnam until 1972, when troops were withdrawn due to increasing political pressure at home. At the same time the progressive withdrawal of US military support saw the South Vietnamese Army left as the sole force opposing the Viet Cong, which proved a defeat in the making.

During the ten years of the Vietnam conflict, Australia provided nearly 60000 infantry, air force and naval personnel. More than 3000 of this number became casualties, among them some 521 combat deaths. This was the cause of intense citizen agitation, expressed in major demonstrations and protests. It was a further manifestation of distaste for the Vietnam War that returning veterans were inexcusably subjected to abuse and hostility. The veterans had only done their country's bidding, and the cruel reaction was completely unjustified. It took the best part of twenty years for this public injustice to be rectified and for the Vietnam service men and women to receive a proper and reasonable level of recognition in the 1987 Vietnam veterans' parade through the streets of Sydney.

Communications in the Vietnam War

Following the decision to send Australian troops to Vietnam, the radio communications equipment required for their support was subjected to a careful review that resulted in the purchase of new equipment compatible with radios used by the US Army.

Of particular interest was the AN/PRC-25, a short-range VHF FM backpack radio which operated at around 40 megahertz, as had its predecessor, the BC-1000. A major difference was that the AN/PRC-25 no longer used valves exclusively. Instead, it used a number of transistors that were able to provide for all of its functions other than the output stage, which retained a power output valve. This meant that the radio's portable power requirements were considerably more modest than in earlier valve equipment, although the output valve still demanded a good deal more battery power than would later all solid-state equipment.

AN/PRC-25 VHF FM (PRJ)

As a complement to the backpack radios, AN/VRC-12 base and vehicle-mounted radios were also ordered, allowing contact to be maintained with platoons and squads some distance away from base.

Base camp radios were accommodated in prefabricated structures that allowed related equipment to be set up, well protected from the weather and frequent rainfall. These AN/MRC-69 relay shelters enabled interconnection between radio resources and telephone and teletype wired links, as well as providing a base where a high-gain VHF antenna could be deployed.

For longer range communications tasks, including providing links to Australia, AN/TRC-75 HF radios were obtained. These suffered from considerable teething problems as they had been modified to allow dual channel (Duplex) operation, unlike the

AN/PRC-25 on patrol (PRO)

AN/PRC-75 in Land Rover truck (PRO)

single channel (Simplex) operation of the original AN/TRC-75 HF. Absence of the relevant technical instruction handbooks also created severe problems initially, when the correct operation had to be determined by trial and error.

AN/PRC-47 HF transceiver (PRJ)

At a later stage, when operations moved beyond the effective range of VHF equipment of approximately 40 kilometres, a new type of portable HF radio was distributed. This was the AN/PRC-47, which allowed the range of fighting patrols to exceed the VHF limit and was not constrained to 'line of sight' operations. Because this equipment operated as a single sideband (SSB) radio rather than using amplitude modulation (AM), which involves the generation of a modulated carrier wave, it was considerably more economical in its power demands. However, the weight of the lead acid batteries that provided power for these radios considerably limited their portability; the nickel cadmium batteries used by the US Army were significantly lighter. The Australian Army also obtained AN/GRC-106 vehicle-mounted HF radios, which were much larger and heavier and not intended for man-portability.

For long-distance patrols involving the Special Air Service (SAS), the A510 (previously used in Malaya) was employed, but was soon replaced by the AN/PRC-64, an all solid-state transceiver made by the US company Delco. It proved very popular given its low power consumption and relatively substantial power output at HF, about 5 watts CW, and was used almost exclusively in this mode to send Morse code, although AM was also available. The AN/PRC-64 radio was used in conjunction with the AN/PRC-47, and was considerably more compact than the other radio used by the SAS, the AN/PRC-25. This VHF radio was generally used in relatively close proximity to headquarters at Nui Dat, but only where VHF communication was physically possible in the absence of obstructions to the signal path.

Delco PRC-64A HF transceiver (PRJ)

Later in the Vietnam War, a new all transistor-based VHF radio, which superficially was identical to the AN/PRC-25, was issued. This was the AN/PRC-77, which had the great advantage of a significantly smaller demand for battery power with the omission of the AN/PRC-25's valve filament heater. The new radio also had the advantage that it could be used with a voice encryption device, the KY-38, which prevented eavesdropping by the Viet Cong.

Late in the war the US Army took delivery of a small two-part communication system, the VHF FM hand-held transmitter and helmet-mounted receiver pair PRT-4 and PRR-9, operating in the 47 to 57 megahertz range. The transmitter produced a mere 200 milliwatts of RF

AN/PRC-77 solid state (PRJ) PRR-9 and PRT-4 VHF FM (PRJ)

power into a short whip antenna, and a built-in microphone allowed a squad commander to issue orders in a normal speaking voice rather than having to shout over the din of battle. The helmet-mounted receiver provided audio output to a flat horn-shaped loudspeaker (actually a 'soft' speaker) or via an earphone. Both parts were supplied with batteries forming part of the equipment. While this device proved much less successful than had been hoped, it can be seen as paving the way to the later personal radios that allowed the intended level of close vocal interaction between members of a squad and the commanding officer, and obviating the use of hand signals.

The AN/PRC-64, the replacement for the A510, had the advantage of light weight coupled with low power demands, being fully transistorised. On the other hand, it was a crystal-controlled device and somewhat limited in its frequency agility. Towards the end of the war, the increasing use of SSB modulation, with its greater efficiency and corresponding reduction in power requirements, suggested that a new type of radio was required for use at HF and for portable long-range patrol operations. The result was a fully synthesised solid-state SSB transceiver designed by AWA and known as the PRC-F1.

Able to operate between 2 and 12 megahertz, with switch-controlled frequency selection, this relatively lightweight and convenient radio was

PRC-F1 HF SSB transceiver by AWA (PRJ)

very simple to set up and operate. Able to produce 10 watts peak envelope power (PEP) in SSB mode or 5 watts of CW for Morse code, here was a very worthy successor to the solid-state AN/PRC-64. Fully sealed against moisture ingress, the PRC-F1 had been developed specifically for use in tropical jungle environments and was powered by a battery carried in a case attached below the main chassis housing. The case was constructed of magnesium alloy, which resulted in a reduced load to be carried coupled with excellent damage resistance.

Backpack radio PRC-F1 in the field (RAS)

14 The Solid-state Revolution

At the dawn of the wireless revolution in the late nineteenth century, many methods of detecting radio frequency energy were experimented with, achieving greater or lesser success. Initially the coherer, first developed by Edouard Branly and Oliver Lodge, was refined by Guglielmo Marconi and incorporated into his wireless telegraphic system. The coherer was a temperamental and unreliable device even in its most refined form and better means of detecting radio energy were soon discovered and applied.

Cat's whisker detector (PRJ)

Most important of these various devices was the mineral or crystal detector. Perhaps the best-known version was the cat's whisker detector, in which a crystal of galena (silver lead) was touched by a fine steel wire which rectified the radio frequency waves and allowed them to be converted into audio frequency signals, audible in a set of earphones.

The invention of the thermionic valve by Professor J Ambrose Fleming in Great Britain in 1904, and its subsequent development as the triode by Dr Lee De Forest in the United States, resulted in a device that was capable of amplifying low-level radio frequency signals, and the crystal detector fell into disuse. Some years later, in the early 1920s, it saw a short revival as an inexpensive means of receiving the early radio broadcasts—but as valves were introduced into consumer radio receivers crystal detectors were relegated to the activities of youthful radio experimenters.

Solid-state Rectification and Amplification

During the late 1930s, the problem of valve amplification in telephone repeater services and the general long-term unreliability of the valve led to the search for an alternative. The research director of Bell Laboratories, Mervin Kelly, had observed the ability of copper oxide to rectify alternating current as applied in early power supplies for valved radios. This made him wonder if it might be possible to discover a solid-state rectifier to replace the thermionic valve rectifiers then in common use. Working with him then, and again later, shortly after the end of the Second World War, was a young scientist who had become familiar with the mysteries of quantum mechanics. This was William Shockley, a PhD student from the Massachusetts Institute of Technology (MIT).

During the Second World War, it had been discovered that conventional thermionic vacuum tube diodes were intrinsically insensitive to the very high frequency radio signals reflected from a target in a radar system. In the search for an alternative the humble cat's whisker was found to exhibit high sensitivity at very high frequencies. This led to the creation of a new generation of stable and sensitive crystal detectors that found their way into many Allied radar receiving systems and later into the crystal radios of youthful experimenters.

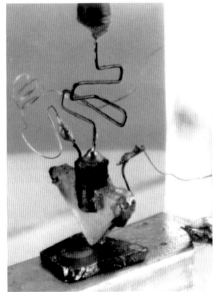

Point-contact transistor (INT)

This development, coupled with his earlier observation of copper oxide rectification, induced Kelly in 1945 to set up a team of experimenters, which Shockley was given the task of leading. In only a couple of years the group was able to construct a crude amplifying device similar in principle to the cat's whisker detector but with the addition of a third electrode. Given the name 'transistor', this new solid-state amplifier initiated a technological revolution in which the ubiquitous valve was made obsolete in only a few years, except in cases such as broadcasting stations where very high-power radio frequency energy was required.

The first-generation point-contact transistors shared the same defects as the cat's whisker detector—fragility and instability. Shockley's work to overcome these problems resulted in the junction transistor, based on a thin sheet of highly refined and doped silicon with contacts on its front and rear faces and one edge. The junction transistor lies at the heart of virtually all subsequent developments in solid-state devices up to the present.

Acceptance of the new device was initially slow, because by that time the valve had been highly developed and reduced in size so that compact and highly reliable equipment had become possible. Not only in the fields of telecommunications and telephony were valves providing highly satisfactory service, but also in an entirely new technology related to computation, the electronic computer.

The Electronic Computer

Valves had been employed in the first major computational machine that only much later was described as a computer, the Colossus of Bletchley Park, which straddled the world of codebreaking but was to remain a secret for many years. For the best part of a decade after 1945, computer developments depended on the reliability of the highly refined valve as a new industry sprang up, in part derivative of the secret machine.

During the later years of the Second World War, one of the first generation of valve-based machines was developed in the United States by J Presper Eckert and John Mauchley, in the form of the massive computer named ENIAC. This had a complement of 18 000 valves and was demonstrated to the public in 1945.

In Great Britain, immediately after the end of the war, Alan Turing, a Bletchley Park veteran then at the National Physical Laboratory, was involved in creating a computer known as the Pilot Ace. At Manchester University, another veteran of Bletchley Park, Max Newman, was engaged in the creation of the Manchester Mark 1 computer, also known as the Manchester 'Baby'.

In Australia in 1947, under the auspices of the Commonwealth Scientific Industrial Research Organisation (CSIRO) a computer development project was commenced under the direction of Trevor Pearcey and Maston Beard. The CSIR Mark 1 computer, also referred to as CSIRAC, was commissioned at the CSIRO Radio Physics Laboratory in Sydney in 1952 and operated there until transferred to the University of Melbourne in 1956. There it was operated for a further eight years until decommissioned in 1964 with the introduction of a new computer imported from the United States, the Control Data 3200.

In 1952, there appeared to be significant prospects for the development of a computer industry in Australia, but the industrial might of the United States won out. Instead, a steady stream of computers arrived from the United States, and from Great Britain, and purchasing replaced manufacturing.

Solid-state Integration

The progressive introduction of the transistor as a substitute for the valve had an enormous impact on two specific aspects of electronic equipment after about 1955. Firstly, the transistor was significantly smaller that even the most compact valve but secondly, and perhaps more importantly, by eliminating the need for a heater to release electrons into a vacuum space, the transistor consumed far less power. These characteristics were very early appreciated by suppliers of military telecommunications equipment. The rapid introduction of field radios with extended operating capacities made possible by the reduced demand on battery power was an early manifestation of this change; soon the heightened level of reliability of transistors also became apparent.

The reduced heat load imposed on associated components in radios and other electronic equipment was clearly demonstrated in the reduction of component failures. In the civilian world, this effectively destroyed the television service industry. Solid-state televisions had an extremely low failure rate whereas their valve-using predecessors had failed regularly and required skilled attention to keep them in operation.

In the military arena, reduced size and weight were highly desirable, and the search for new techniques drove the solid-state industry to ever greater efforts. By the 1960s, not only were military demands pressing the search for smaller systems but the burgeoning computer industry had also developed an insatiable appetite for solid-state devices used as digital switches on logic boards and elsewhere.

When engineer Jack Kilby at Texas Instruments discovered that more than one transistor could be mounted on an understratum of silicon, his patent spawned the rapid development of integrated circuit technology, and devices incorporating it are now found in almost every area of the electronics industry. The effect of the integrated circuit on computer design is an example of what miniaturisation made possible. The logic board built with an early transistor–transistor logic (TTL) array was conventionally the size of half an A4 sheet of paper. Using an integrated circuit, this same device was reduced to the size of a postage stamp.

The creation of the integrated circuit with the associated dramatic reductions in size and power demand allowed the development of the minicomputer and opened the way towards a new breed of miniature or microcomputers. In the mid-1970s the collaborative efforts of a pair of young experimenters in a garage in Silicon Valley in California led to the creation of the twentieth century's best-known microcomputer, the Apple. The young men were Steve Jobs and Steve Wozniak and the company they founded in 1976 remains at the forefront of innovative microcomputer design, smart telephones and associated miniature computing devices.

The story of the first Apple computer has been told repeatedly as well as that of its immediate predecessor, the Altair 8800. The impact of these early microcomputers is very hard to overstate, given the time at which they were developed. Their story in relation to military communications is covered in Part 4.

This new species of computing machine, the microcomputer or personal computer, the PC, in company with the Internet has created a world revolution in technology and communications.

Networks and Data Communications

The appearance of a new voice from space emanating from the radio transmitter in *Sputnik 1* led to the creation of the ARPANET defensive network in the United States. This interlinking of a number of computers across the continent was only possible with progressive developments in methods of connecting computers together to exchange data and messages. The work that was put into connecting computers via a network of links has had the most extraordinary repercussions for business, military and private users.

Before computer-to-computer connections were even contemplated, the long history of electric signalling led back to the early work of Samuel Morse. Employing the switching of direct current generated by a bank of local cells, his simple system provided a means of relatively long-distance communication for the best part of eighty years. It was progressively displaced during the 1920s by machine-generated teletype using Baudot code, in which marks and spaces replaced the dots and dashes of Morse. Morse code did not totally disappear until the mid-1990s, being used by ships at sea until it was forcibly made redundant by satellite-based emergency systems becoming a required facility.

The telegraph service of the early years, operating in the trans-oceanic undersea cable network, was effectively a digital system in which positive and negative voltage swings relative to the earth conductor, sea-water, represented the dots and dashes of the Morse code. This partially overcame the long delays and pulse distortion associated with extreme cable length and absence of amplification, a problem that was not eliminated until 1957 when undersea telephone cables provided the first trans-Atlantic telephone service.

The slowness of the transfer rate of the early telegraph cable system was compounded by its limited capacity—only one operator could send at any one time. Methods of increasing the number of users on the line were achieved through multiplexing, initially by increasing the number of lines from two to four. More effective were rotating commutator multiplex devices in which, by chopping the signal path into a series of very short time segments at both ends of a cable link, a number of connections could be accommodated on a single cable. Later again, it was realised that a significantly greater number of users could be accommodated on a single wire by using a number of radio frequency carriers modulated by audio frequency signals, creating a multiplicity of signals on the same line that could be demodulated at the receiving end, and the initial digital system of linking cables disappeared in favour of a far more efficient system of analogue carrier-borne radio signals. Finally the telegraph was phased out as faster and higher capacity data connections replaced cable connections.

A number of the vital elements of modern-day computer networking had their beginnings in the terrestrial and undersea telegraph and telephone service, among them the electro-mechanical teletypewriter that became an essential interface with early computers—an example appears in the illustration overleaf of the replica of Colossus. The vigorous carriage return of the teletype machine caused the device to 'walk' its supporting bench across the floor, and a very low-tech length of rope was used to keep it in position.

By the late 1950s and early 1960s, the telephone cable system was being called on to provide links between computers geographically well separated—and immediately a basic problem appeared. Computers produced digital signals—on and off pulses—which required conversion to a form able to be conveyed on the analogue telephone carrier service. The problem was solved by a device that was able to produce and receive analogue tone signals to represent the on–off–mark–space digital signal of the computer. The process of modulation and demodulation gave rise to the name 'modem' for the new device.

The first modem was a relatively crude system in which an acoustic

Colossus replica at Bletchley Park (PRJ)

Acoustic modem (INT)

connection was made to a conventional telephone handset with a box that produced tones to replicate the digital input and output signals. An acoustic housing matched the microphone and earphone of the headset with a small loudspeaker and microphone mounted on the modem.

However, as discussed in Chapter 18, this method of data transfer was found to be quite inefficient when employed over the conventional copper cables of the telephone service. As inter-computer networking expanded exponentially, much faster methods of data transfer became essential, and developments in this field continue. These civilian developments have also been relevant in the military context, as a means of allowing transmission of high-grade encrypted data, and to a considerable extent have derived from research funded by government and military expenditure.

15 Technological Change: From Valves to Transistors and Integrated Circuits

In the period 1950 to 1965, military demands for lighter and less power-hungry portable communications equipment saw the progressive reduction in size of radio apparatus coupled with increasing capability and batteries of lighter weight. The introduction of the transistor and later the integrated circuit enabled the creation of apparatus of increasing complexity and sophistication. No more obvious example of this process was the progressive introduction of single sideband modulation to supersede the power- and spectrum-wasteful method of amplitude modulation, still in use in public broadcasting.

Miniaturisation and low power demands found their most sophisticated expression in radios designed for military and clandestine field operations. Some examples from the 1950s and 1960s serve to show the changes that occurred as solid-state technology progressively displaced vacuum tube devices.

Wireless Set A510

The A510 radio, developed by AWA in the early 1950s and used in Korea and Vietnam, was supplied in two cast aluminium boxes which also housed the battery supplies, with the two parts connected by a cable. While possessing the merit of making the signaller who carried it less conspicuous as a target for enemy sniping, the arrangement added to the complexity of the device. This is readily apparent from the two following circuit diagrams.

The A510 had a superhet receiver with reflexed audio and a beat frequency oscillator and a separate transmitter connected to it. This enabled the transmission and reception of Morse code with a continuous wave radio signal for which a miniature key was provided. The valves used in this radio were the miniature battery-powered devices developed in the United States just prior to the Second World War and used in the first generation of consumer broadcast receivers after 1945. The transmitter was crystal controlled, providing both CW and AM voice communications, and operated on frequencies between 2 and 10 megahertz. This range enabled the A510 to take advantage of ionospheric reflections to achieve long-range communications far beyond the groundwave footprint that would otherwise apply.

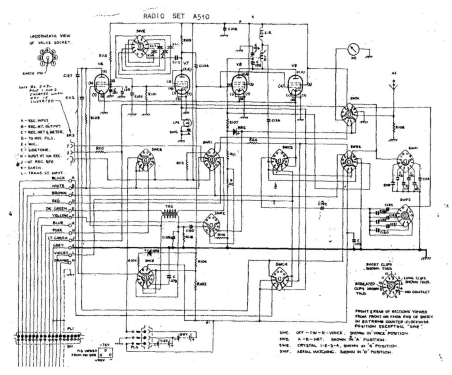

Wireless Set A510: transmitter section (AMP). Larger version p. 336.

The A510 used conventional low power consumption battery-powered miniature valves in both receiver and transmitter, as listed here:

Transmitter valves
1 x modulator 1T4 (V6)
1 x crystal oscillator 3A4 (V7)
2 x power 3A4 (V8–V9)

Receiver valves
1 x RF amplifier 1T4 (V1)
1 x converter 1R5 (V2)
1 x IF 1T4 (V3)
1 x IF AF amp 1T4 (V4)
1 x demod BFO 1S5 (V5)

Technological Change: From Valves to Transistors and Integrated Circuits 231

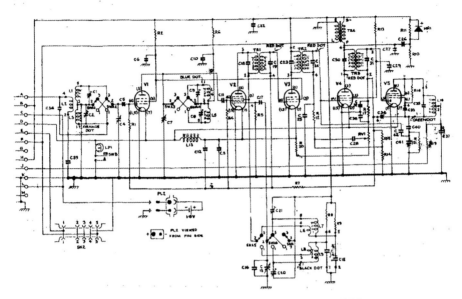

Wireless Set A510: receiver section (AMP). Larger version p. 337.

AN/PRC-64A High Frequency Transmitter Receiver

The AN/PRC-64A radio, initially known as the Delco 5300, and produced during the early 1960s by the US company Delco for the CIA, was tested in Vietnam with very satisfactory results. Being fully transistorised, light and compact, as well as simple to use, the Australian Army found it an appropriate replacement for the A510, and it was adopted for long-range patrol use by the Special Air Service (SAS) after 1970 with considerable success. The AN/PRC-64A appears to be far less familiar in the United States than in Australia.

A 1965 US report into the characteristics of the earliest version of this radio indicated that it was a very useful communications device. The only significant criticism related to the Morse key, which was judged to be too light in construction and precluded the sending of clean signals. (This defect appears to have been corrected in the radios supplied to the Australian Army as the key is identical to that supplied with the A510, a very well made and robust miniature key.)

Among the findings of the US report were:

- The physical characteristics of the AN/PRC-64 are excellent. The radio is small and lightweight and can easily be carried, installed and operated

by one man.
- The AN/PRC-64 is easy to operate. Approximately one half-hour of instruction is required to completely familiarise a qualified radio operator on the tuning and operating procedures of the radio.
- Antenna height and orientation are not critical factors and greatly facilitate rapid employment.

The report concluded that the AN/PRC-64 met all environmental requirements for a patrol radio and was able to provide reliable CW communication at distances between 60–500 kilometres. The frequency range was between 2.2 and 6 megahertz, with crystal control, and the radio was able to generate a RF power of 5 watts. In addition to manual keying, the radio could also be keyed at 300 words per minute with a high speed external device, designed to limit the opportunities to intercept messages by opposing forces.

As can be seen from the three schematics provided here, although the AN/PRC-64A was an entirely solid-state design it bore more than a passing resemblance to the A510.

In the transmitter, the crystal oscillator stage is coupled to an RF

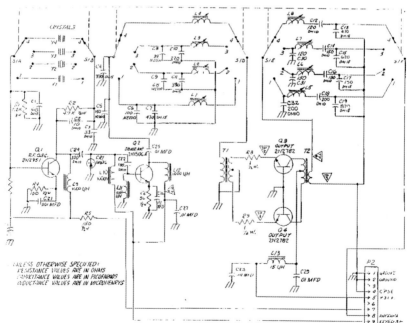

Radio AN/PRC-64A high frequency transmitter board (AMP). Larger version p. 338.

Technological Change: From Valves to Transistors and Integrated Circuits 233

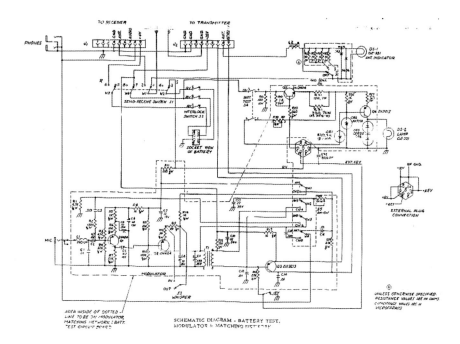

Radio AN/PRC-64A audio frequency modulator board (AMP). Larger version p. 339.

amplification stage which in turn drives a pair of 2N2782 power transistors operating in push–pull configuration to give an output of 5 watts in CW mode. The audio modulator board used to drive this RF transmitter stage is shown in the second schematic.

The battery provides three voltages and comes as a single block that drops into a compartment immediately below the hinged lid of the radio's aluminium container. This lid has a rubber gasket around its lip and can be clamped down onto the container by means of a compression lock so that water is excluded even when the radio is fully immersed.

The ten years that separate the designs of the A510 and the AN/PRC-64A make quite clear the main advantages of substituting transistors for valves: reduction in size and weight and far more economical demands on battery-supplied power. Surprisingly, the AN/PRC-64A is not markedly smaller than the much older valve-based Paraset. A significant difference appears in the receiver design, which in the AN/PRC-64A is a superhet and in the Paraset was a very simple regenerative detector audio amplifier which resulted in very dubious and dangerous operational characteristics.

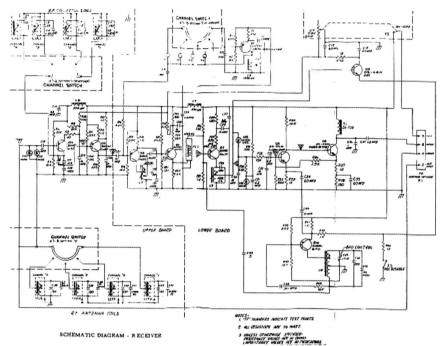

Radio AN/PRC-64A receiver board (AMP). Larger version p. 340.

PRC-F1 High Frequency Transceiver

Developed by AWA in the early 1970s as a transistorised replacement for the A510, the PRC-F1 was initially designated the A512. Unlike its predecessors (including the AN/PRC-64A), this new manpack radio employed frequency synthesis rather than crystal frequency control. Rotary switches allowed the selection of some 10 000 channels, separated by 1 kilohertz steps and ranging from 2 to 12 megahertz. This provided a high degree of operational flexibility and interoperability with a number of existing HF radios. RF power output was selectable between 1 and 10 watts, and could be provided as AM or SSB on the upper sideband. The PRC-F1 could also produce CW for communications in encoded Morse code.

Technological Change: From Valves to Transistors and Integrated Circuits 235

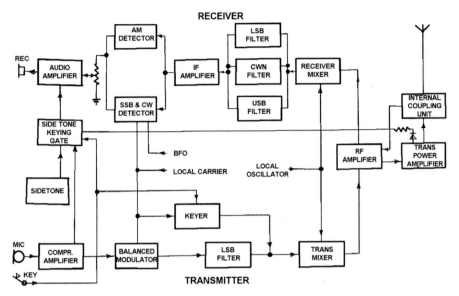

PRC-F1 block diagram (AMP). Larger version p. 341.

AN/PRT-4 and AN/PRR-9 Tactical Squad VHF FM Radios

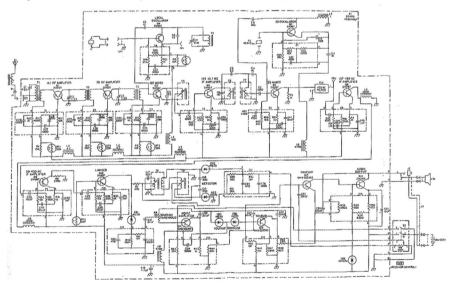

AN/PRR-9 VHF FM receiver, helmet mounted (PRJ). Larger version p. 342.

Although a rather less than successful attempt to provide front-line troops with a small, light means of radio communications, the early US transmitter and receiver pair AN/PRT-4 and AN/PRR-9 (see Chapter 13) are of considerable interest in a technical sense. Interestingly, in physical appearance these radios bear a considerable similarity to much more recent tactical squad radios issued to the Australian Army. The accompanying schematics show that they were fully solid state and employed second-generation transistors to achieve communication, using FM at VHF.

The fundamental problem with this transmitter–receiver pair was that it could not provide two-way communication. The squad commander was able to issue orders to troops, but there was no facility to respond other than by the traditional hand signals. It also appears that a major problem of reliability led to rapid displacement by the somewhat earlier and highly reliable AN/PRC-25 backpack radio, which was used down to squad level, although not the AN/PRC-25's original intended target. Nevertheless, the light weight and reduced size of the AN/PRR-9 and AN/PRT-4 pair established a model for later squad radio development that remains influential and relevant.

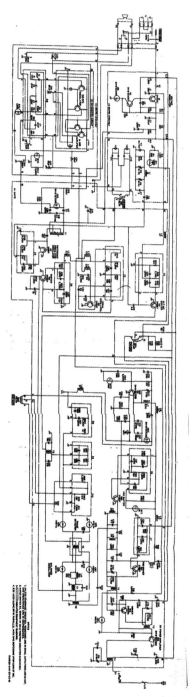

AN/PRT-4 VHF FM transmitter, hand-held (PRJ). Larger version p. 343.

Technological Change: From Valves to Transistors and Integrated Circuits

AN/PRC-77 VHF FM Backpack Radio

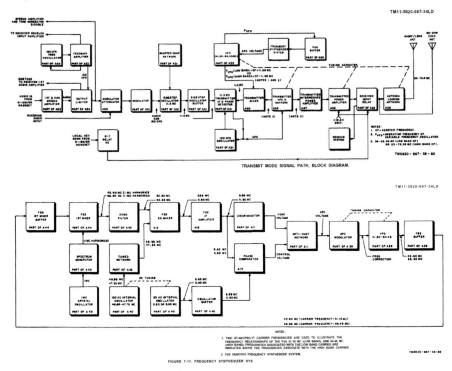

AN/PRC-77 VHF FM backpack radio schematic (AMP). Larger versions pp. 343–44.

Superficially, the AN/PRC-77 VHF FM backpack radio is identical to the AN/PRC-25 which was used extensively in Vietnam by US and Australian troops. Indeed, the only significant difference is the replacement of the valve output stage of the earlier device with solid-state RF power output. Interestingly, the battery supply for the new radio was interchangeable with that of the earlier version, although the valve filament supply was no longer required. Like the AN/PRC-25, the AN/PRC-77 uses frequency synthesis to generate the local oscillation for injection into the modulator mixer, which creates the intermediate frequency.

For the collector of historic radios of the Vietnam War period, distinguishing the AN/PRC-25 from its replacement is virtually impossible from a quick glance. There appear to be no external distinguishing marks, and even when opened up it takes an expert eye to see the differences. In terms of performance, there is also little difference, apart from the power

AN/PRC-77 VHF FM backpack radio (AMP)

hunger of the earlier radio associated with the heater of its valve output stage.

As operators in the battlefields of Vietnam were to discover, however, the single channel operation of the AN/PRC-77 rendered it vulnerable to jamming, and in the absence of voice encryption it was able to be intercepted with potentially disastrous results. Unsurprisingly, it became a favoured object of capture for the Viet Cong and the Northern Vietnamese Army.

These failings were a driving concern in the late 1970s and early 1980s when a device that would be resistant to jamming and interception without requiring separate 'black boxes' to be attached to a radio was being considered by the Australian Army, such as the KY-57 encryption device for the AN/PRC-77. Despite this deficiency, the remarkably reliable and robust AN/PRC-77 remained in service for many years and then was used by school cadet groups. By 2011 it was completely obsolete and was replaced in this training role by the more recent Raven (manpack) and Pintail (hand-held) series of solid-state radios.

The replacement for the unsuccessful AN/PRT-4 and AN/PRR-9 was the AN/PRC-68, a fully synthesised hand-held VHF FM transceiver manufactured by Magnavox in the United States in 1977 and provided to the US Marines in 1980. This radio could be used anywhere between 30 and 79.95 megahertz and over some 920 channels. It could be set up with

Technological Change: From Valves to Transistors and Integrated Circuits

10 preset channels with 50 kilohertz minimum spacing, and with a power output of 1 watt had an operating range of between 500 metres and 2 kilometres depending on which of the two available antennas was used, one long and the other short.

The AN/PRC-68 may well have been observed by the Australian Army and served as stimulus for a hunt for something comparable for use at platoon level. If this is correct, then this ambition took many years to reach fruition, finally appearing in 1996 as the Pintail hand-held radio. In the meanwhile, the long gestation of the Raven series of HF and VHF radios was about to begin.

Point-to-point Wiring and the Printed Circuit Board

Although not necessarily obvious to the casual observer, one of the great electronic revolutions between the 1950s and the 1970s was the general abandonment of point-to-point wiring and its replacement with circuit boards of progressively more complex form. Point-to-point wiring is clearly apparent in the photograph of the under-chassis of a WS No. 101, developed by AWA in the late 1930s.

In the printed circuit board, wiring is provided as thin straps of copper bonded to the surface of a non-conducting substrate (usually epoxy glass

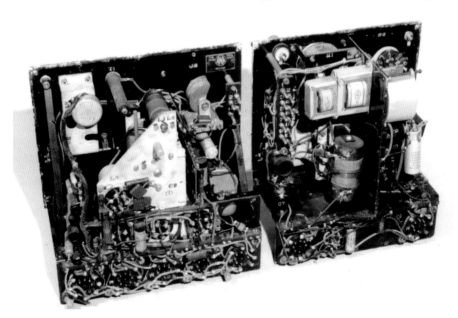

WS No. 101 with point-to-point wiring (PRJ)

fibre). In single-sided boards, the copper base surface is exposed and removed following photolithographic exposure and the application of a chemical etching compound. This allows copper tracks to be formed when the remaining copper surface is removed. Later, when the protective layer over the tracks is washed away in a chemical bath, the copper straps are ready to receive components which can be soldered into position.

To anyone brought up in the era of valves, where the rat's nest of point-to-point wiring was the inevitable problem to be confronted upon delving into any piece of military equipment, the printed circuit wiring board is a complete revelation of neatness and coherent layout. Moreover, it frequently comes close to mirroring the theoretical circuit diagram of the designer and greatly helps comprehension.

Early circuit boards were single-sided, the components being mounted on one side with their connecting leads taken through holes drilled in the board and soldered to the copper tracks on the other side. With increasing circuit complexity, the single-sided board has generally been superseded by double-sided boards in which trackwork on both sides is connected together with holes which are copper-plated and then soldered through. While allowing far more complex component arrangements and wiring, this approach can frequently obscure relationships and make following the circuit quite difficult.

More recently, double-sided boards have been expanded into multi-layer boards, where again connection between layers is effected via plated holes. This type of board is now very much the province of high-component density circuits as used in computers and other associated digital hardware, including smartphones.

What often comes as a considerable surprise is the early date at which the notion of a printed version of conventional wiring arose. One can read about proposals for 'printed wire' as early as 1903, although it was not until the Second World War that a stable, shockproof method of circuit wiring became a pressing need. The new 'proximity' anti-aircraft shells, invented in Great Britain to improve the success rate of anti-aircraft fire, were equipped with a miniature radar carried within the shell casing and set to explode when close to the fuselage of an enemy aircraft, and for this usage shockproof wiring was essential.

Technological Change: From Valves to Transistors and Integrated Circuits 241

The proximity shells proved highly effective against the plague of V-1 pilotless bombs aimed at London in 1944 and 1945 from sites in France. In the extremely stressful environment of high acceleration and rotation in the gun barrel following initial detonation, the circuit board proved highly successful and became the genesis of virtually all subsequent electronic circuit layouts.

Comparisons of the illustrations of the track and component sides of a transceiver PCB and the preceding illustration of point-to-point wiring make it quite clear why the new system became the universal solution to the connection of electronic components in an orderly form that was suitable for mass production and machine assembly.

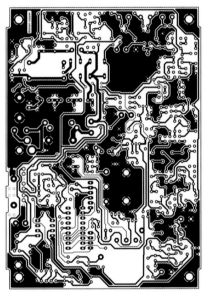
Track side of transceiver PCB (PRJ)

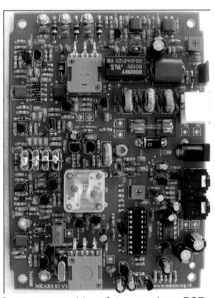
Component side of transceiver PCB (PRJ)

PROJECT 3: Solid-state Double-sideband Transceiver

The most conspicuous change in the technology of military and civilian communications between 1950 and 1975 was the abandonment of valves in favour of increasingly smaller solid-state devices, first discrete transistors and later integrated circuits. This miniaturisation allowed massive increases in the complexity of electronic design without a significant expansion in the overall size of the primary device. In addition, a major increase in the number of power-consuming components could be accommodated without a radically increased demand being placed on the storage battery. The AWA-designed PRC-F1 is a good example of radically improved functional characteristics being provided—without associated massive increases in size and weight.

Duplicating such a radio at the home constructor level, however enthusiastic one might be, is far from straightforward, even with prior experience. Something a good deal simpler than the PRC-F1 is presented as the project for Part 3 to give an impression of what can be achieved with solid-state components and of what might be provided using carrier suppression techniques and sideband transmission, as is now common in military technology.

In November 1994, an interesting project was set out in the electronics magazine *Silicon Chip* involving a low-power (QRP) double-sideband transceiver intended for construction by novice radio amateurs. The article described the ceramic resonator-controlled transceiver and the kit of parts that was available from the electronic supply firm Jaycar. The kit is no longer available, but the design employed common components and even now can be replicated by the enthusiast.

In 1994 this double-sideband transceiver provided an ideal project for the recently formed Sydney Progressive Amateur Radio Club (SPARC) and a number of kits were purchased. The finished results, external and internal, appear in the photographs.

Although relatively unsophisticated compared with the fully synthesised AWA PRC-F1, the kit radio demonstrates some of the key features of the post-valve era. It is both quite small and quite light, as well as being capable of communications over a surprising distance, given its small output power, 1.5 watts peak envelope power (PEP). Moreover, as the schematic on page 244 indicates, the use of discrete transistors rather than integrated circuits makes its mode of operation a good deal easier to understand.

PROJECT 3: Solid-state Double-sideband Transceiver

Despite their utility, the little black boxes of integrated circuit (IC) components, visible in the interior photograph, effectively disguise their internal mode of operation. When they fail, they can only be replaced in their entirety. IC sockets represent a sound insurance policy for this reason.

Solid-state double-sideband QRP transceiver (PRJ)

Double-sideband low power transceiver (PRJ)

Also observable from the circuit diagram, this radio is frequency controlled with a ceramic resonator whose frequency may be varied over a small range by means of a variable capacitance generally referred to as a VXO configuration. This can be useful in allowing the radio to be tuned to avoid competing signals. The resonator serves both as the source of RF for the transmitter, and as the local oscillator for the direct conversion receiver where the incoming signal is mixed with it. Mixing results in an audio frequency output which is then amplified to the level sufficient to drive a loudspeaker, although sensitive earphones are otherwise sufficient.

Although single sideband (SSB) is now common on the military, commercial and amateur HF bands, this radio produces the simpler-to-generate double-sideband (DSB) output. This mode has the particular merit of reduced complexity and component count. DSB shares with SSB the desirable characteristic of a much more economical demand for power than conventional amplitude modulation (AM) as used for radio broadcasting. The two sidebands, lying above and below the suppressed carrier frequency originating with the ceramic resonator, can be received by an SSB receiver. As virtually all amateur SSB radios have the facility to switch sidebands, receiving the output of the DSB transceiver is accomplished by simply switching to one or other sideband depending on local conditions and interference.

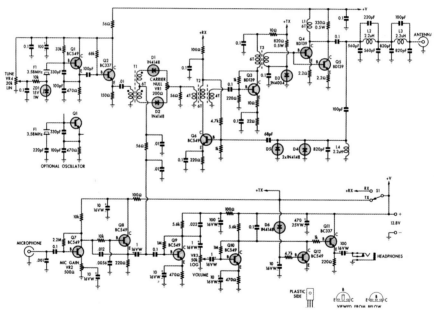

Circuit diagram of low-power double-sideband transceiver (SCM). Larger version p. 345.

Circuit Details

As can be seen from the schematic, although this project was built with a ceramic resonator and operates between 3.5668 and 3.583 megahertz, the transceiver can also be operated with a crystal on some other frequency in the 80 metre radio amateur band. This may conveniently be a 'colour burst' crystal as used in US NTSC television and conventionally has a frequency of 3.579 megahertz.

Frequency stability depends on the 'stiffness' of the power supply, so a Gel-cell lead acid battery is the preferred source of power. Because the transceiver's power output is a mere 1.5 watts, the supply battery does not have to be of a large amp-hour capacity unless protracted contacts are anticipated. For the radio built by SPARC members, 7.2 amp-hour 12 volt Gel-cells seemed quite adequate.

COMPONENT LISTS

Hardware and pots
1 x PC board, code 06110941, 143 x 71 mm
1 Jiffy box, 196 x 112 x 60 mm
1 black binding post
1 red binding post
1 x 3.58 MHz ceramic resonator (F1)
1 x SPDT toggle switch (S1)
1 x 4-pin microphone panel socket
1 square mount SO239 panel socket
1 x 6.5 mm stereo jack socket
1 x 200 Ω horizontal trimpot (VR1)
1 x 500 Ω horizontal trimpot (VR2)
1 x 50 kΩ log potentiometer (VR3)
1 x 20 kΩ linear potentiometer (VR4)
2 knobs
15 x PC stakes
4 x F14 ferrite balun formers (L1, T1, T2, T3)
3 x 2.2 µH RF inductors (L2, L3, L4)

Semiconductors
7 x BC549 NPN transistors (Q1, Q6–Q10, Q12)
2 x BC337 NPN transistors (Q2, Q11)
3 x BD139 NPN transistors (Q3–Q5)
5 x 1N4148 diodes (D1, D2, D4, D5, D6)
1 x 1N4004 diode (D3)
1 x 15 V, 1 W zener diode (ZD1)

Resistors (0.25 W, 1%)
1 x 2.2 mΩ
2 x 1 mΩ
1 x 68 kΩ
1 x 33 kΩ
3 x 10 kΩ
2 x 5.6 kΩ
2 x 4.7 kΩ
3 x 1 kΩ
1 x 820 Ω 0.5 W 5%
3 x 470 Ω
1 x 330 Ω 0.5 W 5%
4 x 220 Ω
1 x 150 Ω
3 x 100 Ω
3 x 56 Ω
1 x 22 Ω
2 x 10 Ω
2 x 2.2 Ω

Capacitors
1 x 470 µF 25 VW electrolytic
3 x 100 µF 16 VW electrolytic
5 x 10 pF 16 VW electrolytic
3 x 1 pF 16 VW electrolytic
13 x 0.1 pF monolithic
1 x .022 pF MKT polyester or greencap
1 x .012 pF greencap
3 x .01 pF ceramic
1 x .0056 pF greencap
1 x .001 µF ceramic
3 x 820 pF ceramic
2 x 560 pF ceramic
1 x 330 pF ceramic
1 x 220 pF ceramic
4 x 100 pF ceramic
1 x 68 pF ceramic

Miscellaneous
Screws, nuts, spacers, medium-duty hook-up wire, shielded cable, scrap aluminium

The zener diode included in the list is used to vary the frequency of the ceramic resonator by varying the voltage applied to and controlled by the pot VR4, which is a 20 kΩ linear potentiometer. Apart from plugging and soldering the components listed into the double-sided PC board, the main chore is the winding of the 2-hole, figure-of-8 Type F14 ferrite inductors using 22 and 26 gauge B&S enamelled copper wire, as follows:

L1	6 turns 22 gauge wire
L2, L3 and L4	2.2 µH RF inductors (discrete, prewound items)
T1	6 turns trifilar 26 gauge wire (see below)
T2 primary	4 turns 26 gauge wire
T2 secondary	4 turns 26 gauge wire
T2 collector for Q6	6 turns 26 gauge wire
T3 primary	6 turns 26 gauge wire
T3 secondary	4 turns 26 gauge wire

The balun formers are 12 x 12 x 7 mm in size and here it is important to understand the definition of a 'turn'. To make a single turn, pass the requisite wire through hole number 1 and out through hole number 2. The wire should then continue around and again be threaded through hole number 1 and out through hole number 2. The two ends should finish up on the same side of the balun former.

The trifilar winding for T1 requires three 400 mm lengths of 26 gauge wire which are twisted together with five twists per centimetre. Keeping the wires parallel, place the three ends in the chuck of a hand-twist drill, and rotate at an even speed. A speed-controlled electric drill could also be used.

A key element of this project is the double-sided circuit board, which is still carried by RCS Radio in Chester Hill, a suburb of Sydney. This organisation has a website on which can be found a listing of all boards produced for the magazine *Silicon Chip* <http://users.tpg.com.au/rcspcb/index0.htm>. The board bears the reference number SC 06110941 and given the double-sided nature of its construction it may be more attractive to purchase it in prefabricated form rather than undertake home fabrication. However, for the intrepid constructor, the artwork for the two-sided board is provided here. The usual processes of circuit board construction involving photo-resist and etching should produce an acceptable result.

PROJECT 3: Solid-state Double-sideband Transceiver

Construction Comments

A maximum null of the suppressed carrier will give the best results. This can be achieved by adjusting the variable resistor, VR1. This process will be assisted if the diodes D1 and D2, which form a part of the balanced modulator circuit, are a matched pair, but careful adjustment of the variable resistor should produce adequate results. This test should be undertaken with the input from the microphone disconnected

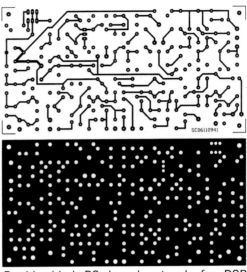

Double-sided PC board artwork for DSB transceiver (SCM)

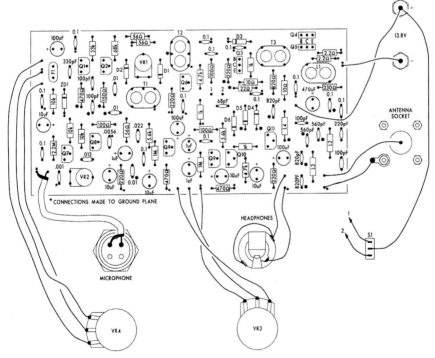

Schematic layout of circuit board and external wiring (SCM). Larger version p. 346.

while the transmitter is powered up and the output monitored with a receiver able to receive the carrier. This may well require the antenna for the receiver to be disconnected and the metal case of the transceiver being positioned to ensure that a signal from the carrier oscillator is not received direct rather than the output stage providing the signal. A short length of hook-up wire placed in the antenna output terminal should ensure that the signal to be detected is derived from the RF power output stage.

Be aware that there are some errors in the original circuit diagram on page 247 that are easily corrected:

- The transistors that sit together at the bottom of the board, marked 9 and 10, should be renumbered 10 and 12.
- The renumbered transistor 12 should have its emitter soldered to the ground plane. It is shown without the asterisk which indicates that such an earth connection is required.
- At the top of the board, slightly left of centre, the asterisk has been omitted from VR1. It should have its bottom leg soldered to the ground plane. Unless this is done, the audio amplifier will give no output at all—which is somewhat discouraging.

To operate this transceiver with more than a matched dipole antenna, an antenna tuning unit is necessary. A quite simple modification of the original circuit allows the production of a carrier for tuning purposes. The additional wiring is shown in the two diagrams here. The first shows the method of unbalancing the balanced modulator so that a carrier is produced. The second shows how the addition of a switch and Morse key operated via a normally closed jack socket enables the production of continuous wave (CW) radio frequency and allows tuning up to be accomplished.

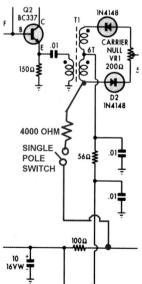

CW changeover switching circuit (PRJ)

The plastic case mentioned in the original *Silicon Chip* article is significantly larger than necessary to house the circuit board and the various external components and sockets. For the author's version, a box that originally held an external floppy disk drive for an Amstrad personal computer was pressed into service. No doubt a metal box could be bent up or some other source

PROJECT 3: Solid-state Double-sideband Transceiver

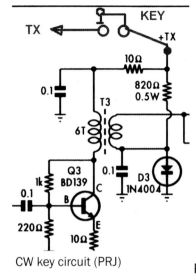

CW key circuit (PRJ)

could provide the requisite case.

Although a very simple transceiver, this project demonstrates in an easily replicated package the most conspicuous features of solid-state design of the period covered in Part 3 of this volume. Light weight, compactness and reduced power consumption were all features of radio apparatus for military use. In the next phase of field radio design, these features became more apparent as integrated circuits allowed ever-increasing complexity. The introduction of microcomputers and their progressive integration with communications devices is discussed in Part 4.

For readers who might be interested in replicating this radio, the November 1994 issue of *Silicon Chip* is still available from the editor at PO Box 139, Collaroy, NSW 2097, Australia.

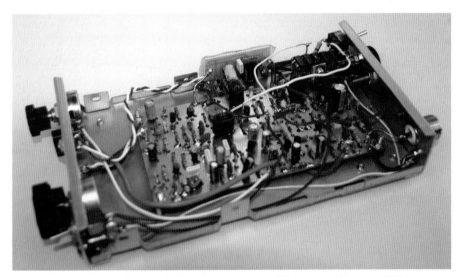

Transceiver showing additional loudspeaker amplifier to middle rear (PRJ)

Part 4: 1976–2012

16 Towards a New Century: Changing Warfare

Since the conclusion of the war in Vietnam in 1973, the complexion of international threats to peace has gone through extensive revisions. Most notable was the removal in December 1991 of the Soviet Union as the source of long-term tension when it dissolved into fifteen post-Soviet states (although today's Russia grows ever more contemptuous of the West). However, the resolution of the Cold War without recourse to the world's vast armoury of nuclear weapons has not produced a cessation of military activity—on the contrary, fighting and military action has been more or less continuous since 1975 and, in more recent times, has been fuelled by religious and sectarian disputes which seem inevitably to lead to recourse to armed conflict. In particular, the Middle East remains a hotbed of mutual antagonisms driven by irreconcilable ethnic and religious differences.

International Warfare and Communications Developments

In the period since 1975, a number of nations have developed nuclear weapons, most worrying being those where conspicuous divisions provide the tinder that could start a new world conflagration. North Korea, with its militant and apparently irrational leadership, is perhaps the most worrying, as it clearly harbours long-term ambitions to regain control of South Korea. In this instance the underlying conviction is communism. Just as potent is the growing capacity of Iran to muster nuclear armaments, despite consistent denial of aggressive ambitions.

During this same period, Australia's involvement in military activities and international disputes has been largely peripheral, derivative of its active membership of the United Nations, the long-standing treaty relationship with the United States, and historical ties to Great Britain. Until quite recently Australian troops in the main were engaged in peacekeeping initiatives sponsored by the United Nations. Since the beginning of the new century, however, such peaceable engagements have developed into significantly more dangerous enterprises as terrorism has become the new face of international unrest. This development has in turn created a new type of asymmetric warfare to which conventional military forces have been constrained to respond and adapt, and has been reflected in requirements for new forms of communications and related equipment.

By the end of the Korean conflict in the 1950s, even in conventional warfare, methods of communication had changed to meet the demands of command and control of aircraft and ground forces to ensure that 'friendly-fire' accidents in particular were avoided. The new hit-and-run tactics of guerrilla fighters lead to ever more frequent opportunities for accidentally misdirected attacks, and the pressure to find better and more compatible radios and associated communications facilities has grown, giving rise to a new generation of more complex (and expensive) apparatus. What also increased after the 1950s was deliberate eavesdropping on military communications, countered by the development of voice encryption and electronic responses to the deliberate 'jamming' of signals.

Asymmetric Warfare

Until 1975, warfare had generally involved conventional, so-called symmetrical conflicts with clearly defined and recognisable opponents. Since that time, warfare has tended to change with the emergence of guerrilla and terrorist forces, frequently involving persons indistinguishable from ordinary civilians.

Since 1975 an increasing number of terrorist-inspired events have taken place, with the most extreme example occurring in September 2001. Few will forget the terrible images of aircraft plunging into the twin towers of the World Trade Center in New York and their ensuing destruction. This act, as unexpected as the Japanese attack on Pearl Harbor in 1941, provoked a passionate reaction in the United States that led to the destruction of the administration of Saddam Hussein in Iraq, and to war in Afghanistan. The war in Afghanistan, which saw the involvement of Australia as an ANZUS ally of the United States, was directed at destroying the terrorists and their supporters, the Taliban, in their home territory.

In 2013, with the United States committed to 'drawing down' its forces in that region, Australia continues to staunchly support a war that remains unlikely to conclude satisfactorily. Given the history of conflict in Afghanistan over the last 150 years, this is unsurprising. In the continuing campaign, the traditional and long-standing Afghan method of guerrilla warfare has forced the Australian military to reconsider in great detail its approach to fighting, including its need for tactical communications for command and control.

Australian Peacekeeping and Warfare Actions

In the ongoing Afghan campaign, fundamentally different approaches to peacekeeping and warfare have been apparent in Australian activities as compared with its allies, and particularly the United States. As in Vietnam, the US military seem firmly wedded to the notion that, in concert with ground-based infantry operations, victory can be achieved with the use of 'stand-off' airborne attacks, employing both conventional aircraft and remotely controlled robotic aerial devices. As in earlier conflicts, the Australian military remains committed to a process in which the 'hearts and minds' of the opponents are to be won over.

The Iraq War: Phase 1, 1991

Following the Iraqi invasion of oil-rich Kuwait in 1990, a United Nations resolution allowed the creation of a multi-national army led by the United States which was given the task of ejecting the invaders. Air attacks begun in January 1991 continued for the next month and a half. After this a ground attack was launched, and in just four days the Iraqi Army was defeated. A ceasefire was ordered on 28 February 1991 and hostilities formally ended in early April.

In this conflict Australia's contribution was almost exclusively naval, in the form of the frigates HMAS *Adelaide* and HMAS *Sydney* and the supply ship HMAS *Westralia*, designated Combined Task Group 627.4. A detachment from 16 Air Defence Regiment was deployed on *Westralia* to provide missile launching capability and to provide air defence to that ship.

Following the ceasefire, and an uprising of northern Kurdish dissidents who saw an opportunity to achieve independence, a small contingent of Australian army personnel was supplied to give logistic, engineering and medical support.

East Timor 1999-2000

During the Second World War, Australian troops had been materially assisted by the indigenous inhabitants of East Timor. This created a national debt that was strongly felt and acknowledged by many ex-service personnel and their descendents. Following the withdrawal of Portugal from its colony in East Timor in November 1975 and the declaration of independence, just nine days later East Timor was annexed by Indonesia.

The majority of the population actively opposed Indonesian colonisation, and in 1999 Indonesia's President Habibie responded to demands for independence by announcing a referendum. The announcement was met by extreme and quite uncontrolled violence by Indonesian supporters in East Timor, which led to the establishment of the United Nations Mission in East Timor (UNAMET) to supervise the referendum.

While the referendum had no doubt been intended to frustrate moves for independence, the outcome was clearly completely unanticipated—overwhelming support for independence. The reaction of the supporters of Indonesia, the militant local militia, was to indulge in increased violence, in which it appears the Indonesian Army both connived and directly supported. Australia, with the support of the United Nations and the United States, was given the task of leading an international peacekeeping force designed to quell the violence. In September 1999, the International Force for East Timor (INTERFET) was deployed under the command of Major-General Peter Cosgrove. This peacekeeping operation, which at times was distinctly warlike in character, overall involved 5500 Australian military personnel and a contingent of 1100 from New Zealand. Support from 22 other nations swelled the INTERFET force to 11 000 at its maximum. East Timor is now a fully independent nation.

The Iraq War: Phase 2, 2003-2011

The destruction of the World Trade Center galvanised the American public into demanding retribution. US President George W Bush interpreted this national fury as a mandate to overcome the forces that had spawned this devastating attack and, as the principal suspect supporter of the al-Qaeda network that had initiated the attack, it was perhaps inevitable that Iraq would become the first target. In 2003, following Saddam Hussein's refusal to permit United Nations weapons inspectors to enter Iraq, a multi-national force with the United States as lead nation was created to force Iraq to comply, with the ultimate objective of ensuring that all anticipated 'weapons of mass destruction' were detected and neutralised. Australia's contribution included naval and military personnel. Once on the ground in Iraq, 'weapons of mass destruction' were never found.

Afghanistan 2001-ongoing

The World Trade Center attack was instigated by the fanatical Islamic entity al-Qaeda, based in Afghanistan and supported by a local fundamentalist group, the Taliban. Al-Qaeda had as its titular commander Osama bin Laden, the son of a wealthy Saudi Arab with close links to the Saudi royal family. From the security of Afghanistan and western Pakistan, bin Laden had been responsible for organising a number of attacks on US installations and organisations before the assault on New York. President George W Bush, son of the President who had launched the previous Iraq war, declared a 'War on Terror' with Afghanistan and al-Qaeda as the principal targets. Eight years later, on 2 May 2011, a military operation ordered by George W Bush's successor, President Barack Obama, led to the death of bin Laden.

The Australian Defence Force contribution to the war in Afghanistan, known as Operation Slipper, has occurred in three phases. The initial phase, commencing in 2001, involved both the Special Air Service Regiment and the Royal Australian Air Force in conjunction with US and other forces, and concluded in 2002 with the apparent defeat of both the Taliban and al-Qaeda. However, like the indestructible Hydra, the removal of one tentacle has seen the generation of others, consisting of determined guerrilla forces.

Following a renewed outbreak of violence in 2005, an Australian Special Forces Task Group was sent to Afghanistan. This included components from both the Special Air Services Regiment and the 4th Battalion, Royal Australian Regiment (Commandos), together with logistical support.

The third and current phase has seen Australian military personnel concentrated in Uruzgan Province, forming part of a Dutch Provincial Reconstruction Task Force attempting to support the capacity of the local Afghan people to maintain a peaceful and democratically led existence. This has involved construction activities and the mentoring of personnel from the newly formed Afghan National Army, an activity that recently has proved somewhat problematic with the deaths of several Australian servicemen at the hands of 'rogue' Afghan soldiers.

Australian military personnel in Afghanistan numbered 1550 in 2009 but now only a small force remains to continue counter-insurgency activities.

17 Military Communication Requirements

In the last thirty years, the nineteenth-century two-dimensional battlefield of Napoleon and Wellington has been completely transformed. Now what is faced by the Australian Army is the modern three-dimensional battlespace in which the additional element of time has also become critical.

During both the Korean and Vietnam wars, as this new three-dimensional battlefield evolved, the problem of adequately controlling dispersed and rapidly moving military land and aerial assets became painfully clear—a problem highlighted by the deficiencies of the available communications systems and apparatus. The interoperability of communications devices assumed major significance when failure to contact and properly direct aerial reinforcements led to disastrous occurrences of casualty and death due to friendly-fire episodes, now referred to more coyly as 'fratricidal' occurrences.

Long-range Command and Control: The Falklands War 1982

Although Australia did not participate, the British campaign to regain the Falkland Islands from Argentine invaders in 1982 occurred at a time when some of the more painful lessons of the Vietnam War were being digested by the Australian Defence Force. The logistical problems with which the British forces had to contend bore considerable similarities to the sort of military engagements that Australia had faced in the past and seems destined to face in the future. The distance between the actual theatre of war in the Falklands and Strategic Command Headquarters in Great Britain mirrored the historical isolation of Australia and the battles it had fought in the past. The Falklands War provided some highly relevant considerations for the development of command and control in which communications are a critical element.

The Argentine army, air force and navy had access to sophisticated, modern weaponry that made them a very difficult adversary in La Guerra de las Malvinas, the 'War of the Malvinas'. That the British eventually overcame the Argentine forces was significantly aided by the availability of effective communications facilities, both in local tactical mode and in the long-range connection to London.

The existence of a special Cabinet Committee, a coherent political body, and its exercise of strategic leadership were key elements in achieving a successful outcome in a campaign conducted over a distance of nearly 13 000 kilometres. The conduct of the campaign in the Falkland Islands theatre, in which local autonomy for the battlefield commanders was carefully preserved, was also of great importance.

In contrast to the problems experienced in Vietnam, communications interoperability was not an issue. However, what did become a major problem was the extent of the access demanded to a very busy communications system, coupled with the volume of information being received that required comprehension and response.

Some years later, US Air Force analysts Colonel SE Anno and Lieutenant-Colonel WE Einspahr noted in a research paper of May 1988:

> the capacity of these modern systems is outpacing the user's ability to sort the information into manageable pieces. Information vital to the conduct of the operation is in danger of being lost within the huge amount of additional information passing across the planning and operational staff's desk. This is particularly so when in any small staff there is always one person who must read and digest every signal and must be aware of all aspects of the operation.

This analysis is just as relevant thirty years on, with the volume of battlefield information having grown enormously. Not only that, the advent of digital data delivery systems via networking means the information is delivered at a much greater rate. Processing information into a useful and comprehensible form is likely to be just as important as delivering the data to the battlefront.

Other important military lessons that came out of the Falkland conflict can be listed as follows:

- The need for preparedness for surprise attack and logistic preparations for such an eventuality.
- The need for planning before the event associated with effective exercises and modelling.

- The need for flexibility to cope with unexpected situations.
- The value of satellite communications.
- The continued value of high frequency communication, given its simplicity and flexibility.
- The need for secure voice communications is a key requirement.
- The need to control the extent of usage of communications facilities and avoidance of overloading of the system.
- The need for three-dimensional communications. Sideways is just as important as up and down.
- The continued need for equipment compatibility and interoperability coupled with adequate capacity.

The Falklands War confirmed that a capacity to 'see what is happening on the other side of the hill' remains just as important an aspect of military communications as it has ever been. Sun Tzu would not be surprised.

Changing Command Structures

During the course of the 1980s the increasing complexities of the battlefield made the conventional vertical chain of command far too slow and cumbersome to allow effective response to elusive and rapidly moving opponents. Stimulated by developments in the US Army, in Australia the solution was seen to lie in the development of digital networks blended with existing radio networks to provide vastly expanded information resources across the battlefield. This led directly to the notion that tactical information needs to be distributed both horizontally and vertically, and this may best be achieved by recourse to what is taken for granted in contemporary business—an information network. What has also become clear is that where a high level of mobility is required, achieving such a network, circumscribed as it is by the realities of physics and the characteristics of radio transmission, is very difficult. The rapidity at which digital data can be distributed in a network is also a direct function of the characteristics of the medium through which it travels, so that the radio frequency spectrum and available bandwidth constitutes the ultimate control.

Knowledge of the geographical disposition of the opponent's forces and accurate information about the character of the terrain are essential functions of any modern military information system. It is imperative that the information is accurate, easily understood and arrives in good time to allow effective planning. The information must also be filtered so that the recipients are not overwhelmed by a mass of indigestible data that confuses rather than enlightens.

In contemporary warfare where unconventional opponents have become the norm, new approaches to command, control and the gaining of intelligence are demanded, especially because the opposition, for all its apparently chaotic approach, has proved remarkably adaptable, relentless and technologically knowledgeable.

Now the use of suicide bombers and improvised explosive devices (IEDs) has generated a war situation every bit as dangerous as anything that has been experienced in the past—one in which communications, intelligence and electronic counter-measures may represent the only truly positive elements in forging a response. Moreover, current military engagements tend to take place in a context of frequent local hostility, in which collateral civilian injuries and deaths generate a persistent background of conflict and controversy.

The future needs of the Australian Defence Force for fundamental battlefield information is a key concern and the basic physical realities of operating communications infrastructure cannot be ignored, particularly where digital networking is the medium over which both digital data and analogue voice information must be carried.

Physical Constraints to Command and Control Communications

The physical constraints to the provision of military communication infrastructure have been elegantly set out in a paper produced by the Land Warfare Studies Centre in Canberra. In Working Paper No. 109 of 2000, analysts Michael Ryan and Michael Frater examined the issues involved in creating a modern system of battlefield command and control. Their conclusions can be seen to underlie current approaches to achieving an adequate system of communications and data distribution, evidenced in the form and content of briefing material currently being issued in requests for tender relating to the commissioning of a battlefield command control and intelligence network for the Australian Defence Force. Ryan and Frater conclude:

> While it is essential that the Tactical Communications System provide a single logical network, it is not possible to provide a single physical network. The range of candidate technologies available to provide access to mobile users constrains the physical architecture to the provision of two major subsystems:

the Tactical Trunk Subsystem and the Combat Radio Subsystem. To extend the range of these two subsystems in dispersed operations, a Tactical Airborne Subsystem is required. Additionally, there is insufficient capacity in the Combat Radio Subsystem (in particular) to cope with the high volume of data transfer required to support real-time situational awareness for commanders of combat forces; this need is met by the Tactical Data Distribution Subsystem. The Local Subsystem simplifies the user interface to the other communications subsystems and the Overlaid Communications Systems.

On the modern battlefield the receipt of timely and accurate information in a form that is easily understood is essential. Moreover, say these authors, without such communications the commander is rendered effectively 'deaf, dumb and blind', surely a prescription for disaster.

The elements of radio communication that come into focus in providing an effective military network are the incompatibility of long-range communications and the need for high capacity to allow data transfer. These requirements can be mutually exclusive, as the higher the data transfer capacity required, the higher is the frequency of operation needed, and high frequency is generally circumscribed by short-range capability. For long-range communications, it is usually necessary to resort to the lower frequencies in the HF band rather than the VHF and UHF bands. It is also a fact that the higher the frequency in use, the smaller the antenna system that can be employed. For this reason, higher frequencies are generally more useful than lower, assuming that range is not a problem.

In the military context, however, and particularly within Australia, long range is a vital element, and reconciling this issue with the problem of data throughput is particularly difficult to resolve. Combat radio networks within a relatively restricted geographical area are able to operate in the VHF spectrum and can accommodate both voice and a higher rate of data flow. For short-range operations where 'line of sight' is not a problem, UHF radio can provide both high-quality voice communications and very high value data flows.

Mobility on the battlefield demands that the military data network also be mobile. This has led to the US Army's introduction of mobile repeater stations for local communications. It has also been suggested that aerial repeater stations—robotic flying machines or helium-filled dirigibles—could be introduced. The use of satellite communications is also a response to the need for long-range communications, with a high level of data flow made possible by the use of UHF links, but this is a very expensive solution, and in a major international conflict is likely to be more vulnerable than HF radio. In this context, it is of interest that the Australian Defence Force has adopted the Israeli Elbit Battle Management System, apparently because it relies on HF communications rather than satellite links.

Military Networks and Communications

In looking at the development of military tactical communications systems and the progressive introduction of networking concepts, a number of basic objectives have applied and will to continue to determine the form of radio and data communication apparatus and its operation in the future. Frater and Ryan summarise these objectives:

- The need for integration of communication elements and connectivity.
- The need for high levels of reliability in equipment and the network developed.
- The need for simplicity of operation.
- The need for ever greater data capacity.
- The need for high-quality voice communications.
- The need for flexibility to meet changing circumstances.
- The need for mobility of both equipment and the network itself.
- The need for security and capacity to prevent interception or interference.
- The need for control of use to avoid overloading the system.
- The need for interoperability and universal connectivity.
- The need for extend range of operations depending on circumstances.

In considering how these objectives are likely to be responded to in the immediate future, it is possible to access information relating to the equipping of the defence forces for the period after 2018. The efforts of the Department of Defence and the Australian Defence Force to replace progressively more obsolete communications and data distribution systems can be appreciated from manufacturers' briefing material placed on the Internet for public consumption. Many of the lessons from earlier combat situations are responded to in the requests for tender, and some even more elementary objectives underlie what is required. In particular, the capacity to share analogue voice and digital data as well as video at all levels of operation on land, in the air and at sea is a fundamental requirement for battlefield operations. Meeting this expectation will demand very high performance, broad bandwidth networks with a range capable of being extended to meet the demands of a particular situation.

Achieving expanded operational reach is likely to require recourse to satellite connections or aerial resources, either mobile or relatively static, using tethered dirigible platforms. On land, the demand for ever more extensive digital data and imagery and video will require the introduction of mobile, vehicle-mounted VHF and UHF repeater nodes, with the ability of individual radios to act as part of a dynamic network providing repeater linking, effectively identical to the peer-to-peer method of file transmission used on the Internet. This will make possible the rapid distribution of large amounts of data in a network in which multiple moving nodes are linked by radio. This would be entirely consistent with the structural notion of the Internet as a linked network of packet-switching nodes, but cast into a new mobile paradigm.

In the face of progressively more demanding requirements, the Australian Army is being equipped with new multifunction radios with extended frequency capabilities, security and frequency-hopping facilities built in. As discussed in more detail in Chapter 20, this has been made possible through the introduction of radically different radio systems in which much of the demodulation and signal processing is carried out in software rather than in hardware. This change has the additional benefit that 'software defined radio' (SDR) is far easier to modify to meet new situations and requirements. Conversely, with less hardware in the 'front end' where radio frequency signal discrimination is most critical, contemporary SDR is potentially vulnerable to narrow band jamming signals, which can seriously overload the front end hardware stages, creating intolerable interference and loss of communications.

18 Technological Change: Digital Development, Encryption, Jamming and SDR

By the mid-1970s, solid-state devices had allowed ever smaller and less power-hungry equipment to be created. However, the large-scale integration of solid-state devices and the development of the microprocessor, in the first generation of mini and microcomputers, would have an even more significant impact on military communications.

Digital Development and Networked Communications

Before the advent of the microprocessor and its application to radio communications, solid-state computers were beginning to impact on military thinking with the development of networked communications. The particular issue that required resolution was how to achieve data transfer over an existing cable network of copper wires that included switching at each end of links originally designed to accommodate analogue voice conversations.

Eventually it was realised that interconnection and the transfer of the digital data computers produced did not require the conventional point-to-point connection used in the standard telephone system. Rather than tying-up a single line for just two users at any given time, digital packages of data could be passed through a network by means of 'packet-switching'. The message could consist of a small string of digital data that formed part of a whole message transferred between computers. If the connection between the computers was via a network of multiple linkages, then the elements of the message, the packets, could go by different routes to the intended destination and be reassembled. This approach had the major benefit that multiple users and multiple packets of data could utilise the network in a given time period virtually simultaneously. The notion of networked computer communication would prove very attractive to the military.

Technological Change: Digital Development, Encryption, Jamming & SDR

Digital Communications

During the Vietnam War, fighting patrols away from their bases relied on short-range VHF FM radio for communications, while wired communications provided the vital link between company bases and headquarters and places less accessible to the enemy. Teleprinters were a major component of the network established on conventional military telephone cable, which also provided a path for voice contacts.

Type TT-4 teletype in operation (PRO)

Right: AN/MTC-1 telephone exchange (PRO)

During the latter part of the Vietnam conflict, apparatus that employed hard-wired connections was installed in prefabricated transportable shelters such as the AN/MTA-4. Servicing equipment in such confined space was a major problem.

The type TT-4 teletype machine illustrated here is operated by an army signaller suitably clad for coping with the heat of Vietnam. This apparatus formed part of the rear HF radio teletype link, flanked by the AN/GRC-106 HF transceiver on the left and a KW 7 enciphering machine on the right. The AN/MTC-1 telephone exchange in the other photograph is set up in prefabricated shelter and operated by national servicemen of youthful appearance.

The Vietnam-era apparatus with hard-wired links was, over the succeeding fifteen years, progressively replaced with equipment designed for digital operations, just as it was in the commercial world. As the capacity for data transfer increased, so too did the demand for facilities that digital data could provide and this too has translated to the battlefield. Typical of the digital display apparatus that became available to the military in the late 1970s and early 1980s are the panel keyboards and displays in the accompanying illustrations.

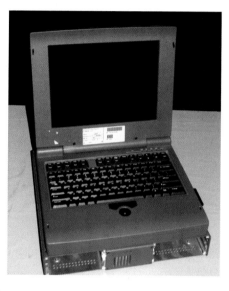

AN/UGC 144 message terminal, 1994 (RAS)

Right: Powerlite 110 rugged field laptop (RAS)

The AN/UGC-144 was described in media material in the early 1990s as a user-friendly 'stand alone' message terminal for composing, editing, transmitting, receiving, monitoring and storing messages. The terminal had a 40 MB internal hard drive and provided single subscriber operation. It was able to operate at a data rate of 45.5 to 32 000 b/s sending and receiving ASCII or Baudot.

The more recent Powerlite 110 unit is physically similar to the Toshiba laptop computer of 1998. However, where the Toshiba used Windows 98 as its operating system, the Powerlite was based on the SPARC central processing unit (CPU), produced by Sun Corporation and using sun4m architecture. With an 85 megahertz microSPARC II CPU and a 250 mm colour screen, together with 32 megabytes of RAM, this was quite a powerful laptop for the period.

The Solid-state Central Processing Unit (CPU)

Discrete solid-state devices such as the transistor made the valve virtually obsolete in a generation, but this was only the beginning of developments in this arena. Of striking importance was the realisation that more than one solid state device could be assembled on a substrate of silicon, from which came the integrated circuit.

The invention of the integrated circuit, and its rapid development, led directly to the emergence of the microprocessor and a new generation of smaller and ever more complex discrete computers, culminating in the present day with smartphones, epitomised by the iPhone and the Android powered units of HTC and Samsung.

In the military context, the microprocessor allowed the creation of small devices to add functions and new capabilities to existing and planned communications equipment. The transmission of text and encrypted voice messages was an immediate target for improvement. By the 1980s, new facilities had improved the utility of military radios and, just as importantly, made them less vulnerable to interception.

Encryption and Avoidance of Interception

From the earliest days of radio's use for military communications, concerns over its vulnerability to eavesdropping had been held. This justifiable concern led almost immediately to the encryption of messages, a tedious and time-consuming chore that was essential if security were to be maintained. Where rapid responses were required, encoding was a significant hindrance to the acceptance of radio, and only slowly was technology able to provide effective solutions.

As the breaking of the Enigma code in World War Two demonstrated, machine- and electromechanically-generated enciphering is susceptible to cryptographic analysis, even where pseudo-random strings of extreme length are generated. The British Colossus demonstrated that a machine could be built to read the output of another machines. Through the work of Alan Turing, this insight was applied to the breaking of the radio-distributed, real-time German telegraphic codes. This was frequently made much easier because of the presence of profound cryptographic errors made by operators and coding clerks during a boring and repetitive task.

In the modern era, where extremely high-speed computation is available through the common desktop and the laptop, apparently impregnable codes generated by machines, even if they be electronic, may prove just as susceptible to analysis and breaking. Moreover, the containment of random code strings in portable communications devices, rather than being provided administratively from a remote location on a frequent basis, may be fraught with danger and, as a matter of principle, should be strongly deprecated. A basic lesson from the two world wars is that, however clever a belligerent assumes itself to be, the ingenuity of the opposing side may match that of the code originator and be quite capable of deciphering the code. This is particularly true now that high-speed electronic computation is universally available and generally accessible to the apparently most unsophisticated adversary.

Landline communications remain generally more secure and harder to intercept than radio communications. Not only is the information contained in radio communications potentially capable of interception and decoding, but its existence also leads to the possibility of obtaining information concerning the location of opponent forces and their disposition and movements. One of the most significant tasks assigned to the French Resistance immediately preceding D-Day on 4 July 1944 was the deliberated and sustained destruction of telephone lines and telephone exchanges, which forced the German Army to rely on radio to distribute orders and information.

In terms of the radio equipment under consideration by the Australian Defence Force, a number of issues need to be confronted. In particular, radio communication is susceptible to disruption by both accidental and deliberate radio frequency interference. During World War One, broad-spectrum signals created conflicts that were just as effective at destroying a radio communications linkage as cutting the wire of a telegraph or telephone system. Technology was only slowly able to respond to this problem. The microprocessor made it possible to introduce methods of avoiding the jamming problem. The most impressive of these techniques, 'frequency hopping', also has benefits in making interception and overhearing far more difficult for the listening enemy.

In the first generation of 'add-on' encryption devices, produced before 1975, the TSEC/KY-57, as applied to the ubiquitous AN/PRC-77, provided secure digitally scrambled voice transmissions. According to military historian Dennis R Starks, this technology was protected by the US military with almost paranoid zeal and denied to Australian forces, which were forced to employ modulated FM to send encoded messages. To do this, a Morse key controlled the audio output of an external oscillator that was used to drive the transmitter—the signal produced in this manner could be heard directly with another AN/PRC-77.

Typical of add-on encryption devices was the MA-4463, made by Racal-Comsec in the United Kingdom and used with the venerable AN/PRC-77.

It was included in *Jane's Military Communications* for 1987 and is likely to be in use with that same radio still, given the remarkable longevity of the AN/PRC-77 and its current use in other countries, especially in the Third World—a lasting tribute to the quality of its initial design by RCA, the company that produced the earlier AN/PRC-25, the famous 'Prick 25' of the Vietnam War.

Frequency-hopping techniques have been incorporated in more recent US-designed radio communication devices under the designation Single Channel Ground and Airborne Radio System (SINCGARS). The output frequency of a transmitter and input frequency of a receiver are 'hopped' around the radio frequency spectrum in a pattern that repeats over time and is supplied to each radio in a communications network. The secure distribution of the coded sequence is referred to as providing 'fills' for the radios in the network.

Frequency-hopping falls into a general category of 'spread spectrum' methods of protecting broadband transmission and reception. This method of modulation and transmission was invented in 1941, rather improbably on the face of it, by the glamorous and mathematically talented Austrian Hollywood movie star Hedy Lamarr, and her collaborator, composer George Antheil. Patented in 1942, this remarkable invention lay forgotten for many years, until in the 1960s it was realised that solid-state electronics now made possible a technique that would render radio communications highly resistant to interception and jamming. Frequency hopping is now found in virtually all military tactical radio communications, allowing a very high level of security when coupled with digital encryption, either contained within particular devices or provided in add-on units.

Although the operational details of military spread-spectrum frequency-hopping devices are shrouded in a cloak of almost impenetrable security, the general principles are now well known, to the extent that the radio amateur community is being encouraged to make use of such technology. A good general discussion of the technique was given by Harold E Price, NK6K, a US amateur, and his explanation of the various spread-spectrum modes is instructive:

- Direct sequence: perhaps the best known and most widely applied method of spread spectrum communications. In it, a narrow band carrier is modulated so that the phase of the carrier is suddenly inverted in accordance with a sequence of coding controls. This sequence is generated by a pseudo-random generator and is of fixed length and repeats over time at a rate that is referred to as the 'chipping' rate.

At the receiver, the encoded modulation is recovered by applying the same code sequence to the modulation as received.
- Frequency-hopping: is the preferred military means of defeating jamming and overhearing. The carrier frequency generated by a transmitter is suddenly and repetitively changed in frequency under the control of a sequence of instructions generated by an electronic device, either as an external add-on unit or as part of the internal circuitry of a transceiver. At the receiver, the frequency of reception is synchronised with the transmitter and then changed in accordance with the frequency 'hops' of the transmitter as determined by the same pseudo-random code sequence. This sequence has to be distributed to the receivers in a military net in a secure manner, either by a separate coding process or by administrative transfer, and is conventionally referred to as the 'fill'.
- Other techniques of spread spectrum include time-hopping, pulsed FM and hybrid systems, the latter involving various combinations of all the other methods of spread spectrum transmission.

In addition to resistance to jamming and overhearing by an opponent, other advantages Price notes include the capacity to allow more than one transmission to exist in the same part of the radio frequency spectrum, as well as a low probability of detection. This latter characteristic relates directly to the relatively low level of the signal, given its dispersed nature across a wide sector of the radio frequency spectrum.

Software Defined Radio
In many respects the most critical technological change in the military sphere since the 1990s has been the introduction of software defined radio (SDR). SDR is the means by which almost all the functions of a communications device are achieved through software control of digital electronics. The significant exception remains the front end where initial reception of the radio frequency signal occurs. Here mixing of the incoming radio frequency signal with a local oscillation produces a signal that can then be digitised and handled by software.

The enormous benefit of SDR is that all the complex functions of a conventional radio, which once were handled by equally complex radio and electronic modules and components, is now undertaken through software. All the conventional operations of demodulation, encoding and frequency movement can be accomplished simply by changing the software, whereas in a conventional radio a range of completely different circuit elements

Technological Change: Digital Development, Encryption, Jamming & SDR

is usually required. This is both difficult and time consuming, as well as exorbitantly costly. For the Australian Defence Force, moving from the hardware excellence of the AN/PRC-77 to new equipment that incorporated provisions for encryption of voice and data, as well as frequency agility, required prodigious effort over a protracted period.

Even the most cursory examination of the elements of SDR reveals that this is a field that quickly develops in programming complexity, although the basic relationships are simple enough. Reference to the literature noted in the bibliography will allow the determined reader to pursue this area of communications technology. SDR has entered the mainstream of radio design, to the extent that radio amateurs are employing it with increasing enthusiasm and despite the problems of getting software to perform as anticipated. Beyond that, in the military context, security has cloaked the software algorithms in the deepest secrecy and only knowledge of the simple principles is available. If any SDR military hardware becomes available in the future as surplus, it can be assumed that any potentially secret elements associated with encryption or frequency hopping will have been meticulously removed. (Some years ago, when the superseded AN/PRC-64A was released to radio amateur and other public purchasers as surplus, even the frequency-determining crystals had been removed.)

The inclusion of a succession of historic radio communications schematics in the earlier parts of this work make it appropriate to provide at least a basic comparison between the last generation of military radios and what is now being developed in SDR. The two diagrams overleaf make the basic differences apparent. The first shows the principal early stages of a superhet radio in which the incoming radio frequency signal is mixed with a local oscillator to produce a difference frequency signal that can be very efficiently amplified before conversion to audio frequency or data in later stages of the radio. The local oscillator can operate as a fixed or variable frequency source and can be generated either by crystal, variable capacity or inductor controlled oscillation, or in more recent times by frequency synthesis with an integrated circuit chip.

In an SDR receiver, the incoming radio frequency signal is converted to an audio form and then digitised before insertion into a special-purpose computer. Here the digital signal can be controlled and manipulated with software to emulate all the common functions of the fully hardware defined radio of previous times.

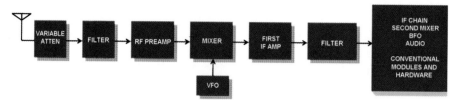

Typical superhet radio showing initial stages and output (RAS)

As can be seen from the SDR schematic, the front end converts the incoming radio frequency signal into two audio signals that are 90 degrees out of phase by means of a quadrature mixer array. These two signal streams are then injected into a digitising module, in the simplest situation a high-quality sound card, and passed to the digital hardware of a contemporary computing engine. Here software control and manipulation take over to allow a flexible response to all the conventional forms of modulation, such as AM, FM, phase-shift keyed ASCII, CW, SSB and so on. In addition, all the complexity of frequency hopping and encryption can be undertaken and adjusted later as situations change.

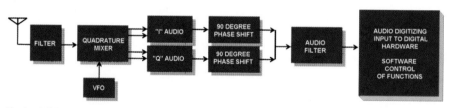

Typical SDR front-end mixer and digitising network (RAS)

As in the conventional superhet radio, the local variable frequency oscillation can be generated by various means; in the most recent radios, this is achieved with a direct digital synthesiser chip or DDS, such as the Si570, which has been employed in radio amateur projects. This particular chip covers a frequency range of 0.1 to 160 megahertz, which is achieved by programming the chip in software.

It seems likely that the development of future military communications will be made significantly easier by the advent of SDR. On the other hand, taking heed of the perception that design development groups tend to produce 'camels' that amble rather than 'horses' that gallop, it may be anticipated that the process of providing the army with new communications systems will be fraught with difficulty and delays and the potential gains from the introduction of SDR may be denied unless determined overall control is applied.

The initiative in the development of communications systems in a digital context has moved firmly into the realms of commercial activity where progress will occur in the conventional way, with the pressures of competition driving rapid responses to changing needs. Consumer demand, the emergence of the Internet, and dynamic interaction and data transfer in a mobile context, have resulted in the creation of equipment which, in functional terms, is what one might expect as a proper response to military requirements.

As discussed in detail in Chapter 20, the military has in recent times turned to off-the-shelf commercially developed communications apparatus (COTS). This is not only because of the impact of ongoing economic strictures but also because the functional characteristics of SDR make such a course of action almost irresistible. Rather than having to meet the costs of prohibitively expensive design and development of hardware that is subject to rapid obsolescence, energy can be expended on software development with its inherent flexibility.

19 Battlefield Communications: 1976–2006

By the end of the 1970s, military signals operations involved a mélange of physically distinct radios and associated apparatus, causing major difficulties for the defence forces in operation and maintenance. However, it was realised that a solution was required, despite the prevailing political distaste for warfare and an associated reduction in military expenditure. Moreover, the political attitudes that had seen support of involvement in the Vietnam conflict had changed in 1972 with the incoming Whitlam government. A new policy of defending the Australian mainland as the primary responsibility of the Army replaced the outward-orientated stance of the Menzies government.

Now any thought of involvement in overseas expeditionary actions received little public or political support, although Australia did find itself called upon to engage in a number of foreign actions and peacekeeping duties. The prevailing political attitude was that no external military threat to Australia's security was likely to occur within the next ten years. The unfortunate result was that necessary improvements to the Army's communications equipment were stalled for the best part of ten years. Indeed, it has been suggested to the author in private correspondence that between 1975 and 1985 there was no capital procurement of new signals equipment, other than expenditures associated with an entirely new range of military communications equipment: Project Raven.

In 1980 the Royal Australian Infantry Corps was operating with radios that would have been familiar to the previous generation of soldiers who had served in Vietnam. The reliable AN/PCR-77 was still in use, operating on VHF with FM over relatively short ranges for infantry intra-platoon communications, as well as for extended links back to company headquarters and beyond.

AN/PRC-77 solid-state FM backpack radio (RAS)

For longer range communications the AN/PRC-47 HF SSB radio was still in use, generally vehicle mounted or used in conjunction with the termination of lines and switchboards located in single portable communications enclosures. For manpack use in the field, the synthesised SSB HF radio described in Chapter 11, the PRC-F1, remained in service. It was also in use as vehicle-mounted equipment, where it was designated the GRC-F2.

AN/PRC-47 HF SSB radio mounted on rear of communications vehicle (RAS)

For ground-to-air liaison, a UHF FM radio was available in the form of the AN/PRC-41. For long-range patrol operations, the AN/PCR 64A was available for AM and CW, with high-speed CW produced by an add-on device pre-loaded with coded messages and sent in rapid bursts of RF.

By 1980, it was apparent to a number of senior signals officers that a radically different design approach was essential if field communications were to be able to meet future military situations. The procurement of a new and different system of integrated radios was needed. Given the prevailing political and public mood, it was also recognised that embarking on such a course of action might be seen as 'raving mad'. Although probably apocryphal, it is rumoured that the codename Project Raven for the next series of Australian military radios grew out of just such an observation. As the reaction in the highest echelons of the Army to a proposal for major reorganisation and rationalisation of military communications and radio equipment was likely to be highly unsympathetic, the 'raving mad' story seems to bear the ring of truth!

PRC-F1 SSB synthesised HF radio by AWA (RAS)

In the early 1980s, when it was finally recognised that the technology of modern warfare made the currently available equipment vulnerable to jamming and interception, the Australian Army began a protracted process of design specification and negotiation. This led to the purchase of replacements for the HF radios, and of equipment that was compatible with a new series of VHF radios that were interoperable with the long-serving AN/PRC-77.

The exercise led to the British firm Plessey, subsequently taken over by Siemens, and in the mid-1980s negotiations with the Department of Defence and the Army produced a new local entity in Siemens Plessey Australia and the production of the Raven series of radios, derived from the Plessey System 4000 family. These radios made use of complementary metal oxide semiconductor (CMOS) technology, employed NSC 800 central processors, and were intended to be highly resistant to jamming through frequency-hopping capability and able to send and receive fully encrypted data and voice.

A new HF radio based on the Plessey System 4000 equipment went into production in 1987. Designated the RT-F100, it was able to interconnect with other, now obsolete, HF radios still in service. This was followed in 1994 by a VHF radio designated the RT-F200 that had a frequency range of 30 to 88 megahertz and was compatible with the AN/PRC-77 still in use in Reserve situations and by school cadet groups.

The Raven backpack radios were intended to form the core of a new combat radio network (CNR) that would enable an expanded and integrated arrangement for command and control across the continent. At

Project Raven RT-F100 HF backpack radio (RAS)

that time Australia was assumed to be the main area of concern and the only theatre of military action anticipated in the foreseeable future. These radios remain in service at the present time, although new radios from Harris Corporation and Raytheon in the United States are currently in the process of acquisition and will entirely replace the Raven series in the next few years.

New techniques of voice encryption and frequency hopping incorporated in the Raven radios presumably made it possible to retain this equipment in service in Afghanistan against a somewhat less technologically sophisticated opponent. However, that situation could change quite rapidly and has led to the Federal Government's decision to undertake the major program of communications enhancement and upgrading discussed in Chapter 20.

Further HF and VHF equipment was provided by the renamed Siemens Plessey Electronics Pty Ltd in the latter half of the 1990s. These radios featured full encryption and frequency-hopping capabilities and allowed army vehicles to be included in the battlefield command and control network being developed. This was referred to as the Wagtail Project and later, as the network was expanded to accommodate digital data, as Project Parakeet.

With the completion of Project Parakeet, the core elements of a new battlefield command support system were in place. The equipment was put into service between 1998 and 2000 and is understood to be in use at the present time.

The Pintail project involved a small hand-held VHF radio supplied by Racal (later absorbed into the Thales Corporation). Designated the RT-F700, it was based on the 1989 Racal AN/PRC-139 handset, featuring three VHF bands and fully interoperable with the Raven VHF RT-F200 radio that was used in vehicles and in the form of an infantry backpack. Pintail radios became available in 1996.

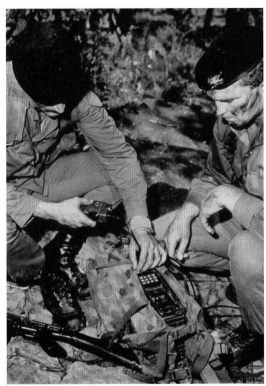
'Filling' the VHF RT-F200 Raven in the field (RAS)

They were later supplemented by the software defined AN/PRC-148, a multiband inter/intra team radio (MBITR), manufactured by Thales.

The hand-held AN/PRC-148 covers a remarkable frequency range of 30 to 512 megahertz with frequency stepping at 5 or 6.25 kilohertz intervals. Frequency-hopping protection is available using a pseudo-random hopping rate, it has a power output of 0.1 to 5 watts, and is able to accommodate data rates up to a maximum of 4800 baud in asynchronous mode or 16 kilobits per second in synchronous mode. Making it compatible with older single-channel radios, the AN/PRC-148 can receive both AM and FM signals as well as data.

In addition to the AN/PRC-148 MBITR, in 2004 a new, highly compact UHF squad radio referred to as the AN/PRC-343, manufactured by SELEX, was introduced. The SELEX organisation is the descendant of the Marconi Company of 1896, created through a series of business mergers that included the English Electric Company (1946) and later the General Electric

AN/PRC-148 MBITR hand-held VHF software defined radio (THA)

Company (1968). GEC acquired Plessey in 1989 and Ferranti in 1990–93 to create a powerful electronics concern involved in communications, radar and defence related technologies. The defence arm of GEC was in 1999 acquired by British Aerospace to form BAE Systems, which in turn created SELEX as a joint venture with the Italian firm Finmeccanica.

Notwithstanding the complexity of its ancestry, the SELEX UHF radio, known in the British Army as the personal role radio (PRR), is both small and lightweight, is simple to operate and incorporates design features that make it highly resistant to jamming and interception. To achieve this desirable character, the PRR uses a spread-spectrum form of modulation and is intended for close-in infantry support, not requiring hand manipulation and so leaving the user free to operate weaponry at the same time.

The PRR communicates on frequencies between 2.4 and 2.483 gigahertz with an operating range of approximately 500 metres. The radio

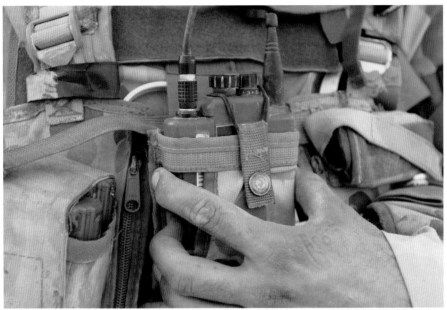

AN/PRC-343 personal role radio (SEL)

SELEX/Marconi personal role radio (SEL)

has a channel capacity of 256, of which 16 are selectable by the user.

The modulation method is known as direct sequence spread spectrum (QPSK) and is associated with continuously variable slope delta modulation (CVSD/CVSDM) voice coding. Power output is 100 milliwatts supplied from two AA primary batteries and the set weighs a mere 1.5 kilograms. The radio is intended to be used by members of a platoon to talk between themselves and with the platoon commander in a fully secure manner.

20   Contemporary Military Communications

Up until 1975, the influence of both Great Britain and the United States was readily apparent in the radio and communications systems used by the Australian Defence Force. A generally robust local manufacturing capability in military and civilian electronic parts and equipment still existed, despite the inflow of inexpensive apparatus from Japan and Southeast Asia and elsewhere. Since then manufacturing capacity has been steadily eroded and a substantial component of defence-related equipment now comes from overseas, a somewhat dismal situation if ever Australia were forced to defend it shores more aggressively than fending off rickety small boats with desperate asylum seekers as cargo. This is particularly obvious when one considers the extent to which contemporary electronic apparatus and communications security is dependent on advanced technology. The silicon chips and processors that make up modern digital equipment today come from overseas, and productive capacity for such devices in Australia is effectively non-existent.

In the light of this downward spiral of dependence on overseas suppliers for all forms of technological apparatus and equipment, it is hardly surprising that military eyes have tended to turn in the direction of civilian off-the-shelf (COTS) military hardware and communications equipment from British and US sources. Nonetheless, after the early 1980s development of the Raven and Wagtail series of radios, a measure of Australian manufacturing involvement was retained in the form of Plessey Siemens Australia.

In an analogous situation, in the replacement of the effective and potent L1A2 self-loading rifle (SLR), in 1988 the Australian Defence Force looked to Europe for a rifle that would fire the US-designed military cartridge used in Vietnam. Where the SLR used the powerful 7.62 x 51 NATO cartridge, a direct descendant of the .303 inch round used in the reign of Queen

Victoria, the new rifle, the AuSteyr F88, based on the Austrian-designed Steyr Mannlicher AUG A1, loads the US-developed 5.56 x 45 mm rimless cartridge (also used in the Colt-manufactured M16 and its derivatives). In the conflict in Afghanistan, the ballistic limitations of this new cartridge have become brutally apparent, while the Taliban continues to employ the .303 cartridge in the highly effective long-range Lee-Enfield No. 1 Mark 3* (SMLE) rifle.

The process of equipping the Australian Army was completed with new rifles manufactured and serviced under licence as the AuSteyr F88 at the home of the SMLE rifle, Australian Defence Industries (ADI) at Lithgow in New South Wales. The Commonwealth Government sold ADI to Thales Electronics Co. in 2006.

Radio Hardware in 2013

More recently, in responding to the need for new communications equipment and facilities, the path to Australian construction and supply has been a good deal more problematic. Overseas supply through local representation, rather than Australian construction and supply, appears to be the preferred solution. The major contractors currently supplying military communications equipment, and battlefield command and control networking facilities, are the Australian distribution subsidiaries of the US technology companies Harris Corporation and Raytheon and the Israeli company Elbit Industries.

A principal contractor for the contemporary production of radios for intra- and inter-company communications is the Harris Corporation with the AN/PRC-152, an SDR hand-held radio that is able to communicate with other equipment in the current deployment of new radios and data sources, as well as having the capacity to operate in conjunction with older, single-signal field radios.

AN/PRC-152 (HAR)

Operating in the VHF and UHF parts of the radio frequency spectrum, the AN/PRC-152 has an expanded array of capabilities and functions and provides them in a remarkably compact and lightweight package. Compared with contemporary smartphones and radio amateur VHF and UHF transceivers, this is no diminutive piece of equipment, although a considerable part of the bulk comes from its long-life lithium-ion battery. The radio weighs just over one kilogram (or 2.4 pounds).

The AN/PRC-152 is a single-channel multi-band radio, covering a frequency range from 30 to 512 megahertz, and has a power output adjustable from 0.25 up to 5 watts. It operates with AM and FM as well as high-speed data for digitised voice and data streams at up to 56 kilobits per second.

As an SDR, it is able to provide encryption of voice and data to a standard approved by the US National Security Agency (NSA), as well as frequency hopping, available via a special Harris proprietary module which is able to be programmed in software. With this capability, the AN/PRC-152 is able to operate in conjunction with older apparatus which depends on encryption systems that are provided both internally or as externally attached modules—even the venerable AN/PRC-77.

The companion to the AN/PRC-152 is the AN/PRC-150 HF packset radio, also produced by the Harris Corporation. This SDR also exhibits a dazzling array of capabilities and is able to traverse a frequency range of 1.6 to 59.999 megahertz with 75 fully programmable frequency slots. It is able to communicate via SSB, compatible AM (SSB with carrier), CW and FM. Power is delivered at 26 volts DC and power output can be set

AN/PRC-150 HF packset radio (HAR)

between 1 and 20 watts. A separate amplifier is available to elevate the power output to 100 watts when the radio is used as vehicle-mounted equipment. The AN/PRC-150 is also able to receive encrypted data at between 9600 and 12 800 bits per second. On VHF, it can operated with frequency shift keying (FSK) at 16 kilobits per second.

As well as a capacity to automatically set up links to the data network and other radios, referred to as automatic link establishment (ALE), frequency hopping is available using serial tone electronic counter-counter measures (ECCM). All these facilities are contained in a backpack radio that weighs a modest 4.7 kilograms—without batteries. Radio frequency energy is broadcast via a whip antenna or various wire antennas, including dipoles.

In creating a modern mobile local network to support frontline troops, Raytheon has reworked a tried and true radio that was introduced well over twenty years ago. This is an interesting reflection of the longevity of solid-state technology on the one hand, and shows the implications of software control on the other. Here modern networking applications have been accommodated without the need for a fundamental redesign or revision of the hardware elements. Where originally the AN/TSQ-158, an enhanced position and location reporting system (EPLRS) radio, was operated in conjunction with a fixed computer server, in its current fourth generation it is able to operate as part of a mobile system utilising an ad hoc wireless network in which all the radios are designed to handle messages as relay points. In this role, the AN/TSQ-158 can be used in conjunction with a separate laptop. Its principal functions are to enable tactical data to be distributed and received, as well as reporting the location of the operator for incorporation into a battlefield mapping system available to other operators.

In addition to these capabilities in the networking role, when used locally the AN/TSQ-158 operator not only has access to all parts of the network but also, as with earlier radios, has protected access to other users via encrypted voice communication and data transfer in frequency-hopping mode. As might be expected of a wireless system that relies on UHF, the local network is limited to 'line of sight' when in terrestrial mode—a shortcoming that may be overcome by means of satellite links or repeaters supported on tethered dirigible balloons. Tropospheric reflection by scattering of RF energy can also significantly extend the range of a UHF network, although this is to a considerable extent subject to the vagaries of the tropospheric environment.

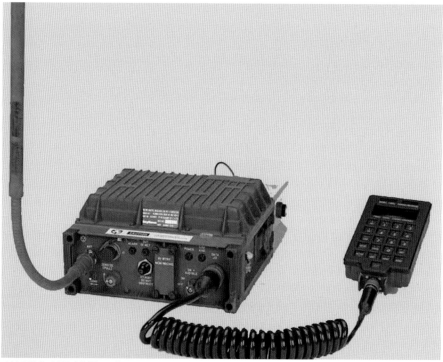

AN/TSQ-158 UHF radio: enhanced position location reporting system (RAY)

The characteristics of the AN/TSQ-158 radio listed here are obtained from Raytheon media material contained in the periodical *Australian Defence Monthly* and referenced in the bibliography:

- Frequency band: UHF, 420–450 MHz
- Frequency hop: 512 times per second over 8 frequencies
- System architecture: synchronous TDMA, frequency and code division multiplexing
- ECM: spread spectrum, frequency-hopping error detection and correction
- Security: embedded crypto, transmission and dual level communication security
- Data rates: two types of low data rate (LDR) needlines with rates to 3840 bps; three types of high data rate (HDR) needlines with rates to 100+ Kbps.
- RS dimensions: 14 x 10 x 5 inches (355 x 250 x 125 mm)

- Prime power: 28 VDC, 16 watts
- Weight: 17 lbs or 26 lbs with batteries (7.7 or 11.8 kg)
- Output power: selectable 100, 20, 3 or 0.4 watts

This radio is intended to be mounted in current army vehicles such as the Bushmaster and M113 armoured vehicles.

In addition to the vehicle-mounted AN/TSQ-158, the Army has been supplied with a companion EPLRS radio known as the Microlight, from the Australian arm of Raytheon. This is a small hand-held radio usually carried in a pouch on the chest or shoulder and able to be operated by voice direct (VOX), allowing the hands to be used for weapons operation. For reception of data, images and maps, the Microlight can be interfaced with a variety of digital devices from laptops to PDAs and smartphones.

Microlight radio for EPLRS (RAY)

Microlight radio for EPLRS (RAY)

Not entirely clear from the literature currently available is how the Microlight differs in operation from the personal role radio (PRR) discussed in the preceding chapter. Given the Microlight's frequency coverage, it may have somewhat better range than the PRR and be less susceptible to masking by intervening obstructions and buildings and thus be more useful in guerrilla warfare and street fighting situations, quite apart from its capacity to receive video and mapping data from the wireless network implemented via the AN/TSQ-158 EPLRS.

21 Future Directions

In looking ahead over the next few decades, and relying on an instinct for change rather than hard facts, it can be asked: What is likely to be the character and requirements of the future infantry soldier and what will his or her needs be in terms of responding to opposing forces? What can be reliably anticipated is that whatever the soldier is confronted with, it is likely to include the totally unexpected and equipment that may not yet even have been conceptualised.

While existing trends in military operations may be expected to give some guidance, the problem with extrapolation is that it is always going to miss the impact of unanticipated, totally new technology. Already we have seen completely new digital devices and network systems invade the battlefield in a remarkably short period. This has occurred in concert with the development of methods of indirect force applied through robotic and aerial armaments and other stand-off weapons and in the context of the emergence of unsophisticated but deadly guerrilla opposition.

Imagining the Future

Over the years Hollywood has been surprisingly prescient in its analysis of future events and technology. One has only to think back to the revolving space station of Stanley Kubrick's 1968 movie *2001: A Space Odyssey* to see the accuracy of its grasp of future possibilities. Even in the much more recent machinations of the *Terminator* quadrilogy, or the over-the-top *Transformers* movies, one glimpses an impression of future battlefields in which intelligent machines, impervious to bullets and explosives, will create havoc and destruction.

However, it can be reasonably expected that in future conflicts there will still be face-to-face engagement. While the cold steel of the bayonet may in general have been relegated to history, high-speed projectiles will persist for the foreseeable future. Their delivery may be more efficient and deadly, as heavier explosive rounds replace inert copper-cased lead, but the inertial impact is likely to remain the key characteristic, just as now. In this scenario, body armour is likely to be of limited use in protecting human flesh. This, it will be remembered, was the reason behind the clanking, lumbering tanks of World War One.

So, with the benefit of media artifice to spur the imagination, one may wonder if *Star Wars* gives a picture of the warrior of the future or is entirely misguided. Already it appears that this movie series was the inspiration

Star Wars fighter (INT)

Future Digger (AMP)

for President Ronald Reagan's demand for a new anti-missile system in space. However, based on illustrations provided by Australian defence boffins, the warriors of the future may appear somewhat more prosaic than their *Star Wars* counterparts, although significantly different to those of the present day.

An Economically Austere Future

With the combination of ongoing economic stringency on the one hand, and a diminishing pool of potential military recruits on the other, it may be predicted in the Australian context that a strictly utilitarian, budget-related army model will prevail in which the infantry will be expected to do progressively far more with existing equipment, or with a limited future palette of new armaments and communications devices. No doubt the political expectation will be that better equipment and better communications will enable limited military resources to fight ever more effectively and with reduced casualties.

The dangers of such a presumption, however, can be seen in the the Korean conflict, where sheer numbers of opponents, prepared to die without qualm, brought the US Army to a stalemate just as awful as that seen in Flanders forty years earlier. Nearly a century before that, hordes of assegai-waving Zulus completely overran and destroyed the flower of the British Army at Isandlwana, and only remarkable tenacity and concentrated rifle fire prevented the same outcome at Rorke's Drift shortly afterwards. Realistically, small professional armies may have a limited capacity to respond to adversaries in huge numbers, even if the most potent modern weapons and air power are available. On the other hand, the availability of the most functionally useful and extensive modern communications equipment and network support might indeed be the element that allows a small Australian Defence Force to respond effectively to massive opposition. We must all live in hope.

Civilian-supplied Communications Systems

The commercial arena is likely to be the supplier of the majority of future military communications equipment. Civilian-supplied communications equipment commonly goes by the abbreviation COTS. What that equipment is likely to involve in a technical sense might appear to involve a security problem, but with software defined radio it is not the apparatus that is important but the way it is programmed. This is reminiscent of the Enigma encoding machine of the Second World War, where the instructions on how to set up individual machines were the crucial element. The discovery of the machine's physical configuration, although certainly exceedingly important, was of secondary moment.

Digital computer-related apparatus together with mobile ad hoc network facilities are likely to be even more important and in demand in the future, leading to their rapid introduction. Facilities of this kind are likely to be used in conjunction with devices deriving from the current generation of small-form communications equipment. Given their simplicity of operation, and the power of the facilities that can be accessed, turning such equipment to military use seems absolutely assured. The only wrinkle in the soldier's use of such devices is that they need to be operated 'hands off' so that armaments receive full attention and manual control. This may well mean that the incorporation of communications into headgear or weapons is the way of the future. The presentation of maps, video and other information in the field of view will need to be achieved without the soldier having to look away from a potential target or environment. This highly desirable facility is already provided in high-speed combat aircraft and in the mundane civilian area, where GPS guidance information is provided on head-up windscreen displays in some luxury motor vehicles. The Microlight device referred to in the previous chapter appears potentially capable of providing the network link required for such a service, and it is only the acoustic and visual display facilities that would need to meet specified criteria.

Local Area Mobile Network Facilities

Potentially the most problematic facility to be demanded on the future battlefield will be elevated data rate networking capacity, able to handle high-speed streamed video imagery in real time along with various graphical presentation services such as mapping and written messages, all in a highly secure environment.

As announced by the successful contractor, Raytheon, the hardware to undertake this onerous task has been selected and is in the process of distribution to relevant areas of the Army. The radio provided for this task is of surprising antiquity in the light of conventional design, manufacture and procurement. However, it has been extensively reworked to provide ad hoc networking capabilities based on a mobile server configuration in which peer-to-peer file handling occurs much as it does in the Internet analogue, except that it is not dependent on static workstations to operate as nodes in the system. Here individual radios act as peers in the hands-on process of file distribution and video streaming.

If this system operates as claimed by Raytheon, then a highly valuable aid to information dissemination will be created with the potential to support the sort of facilities now taken as the norm on the Internet, including mapping, message handling, encrypted voice and positional information derived from GPS satellites. The success of this system will depend on its capacity to overcome deliberate opposition jamming on the one hand, and interception and penetration on the other.

The counterpoint to the Raytheon hardware will be the battlefield command and control software environment, for which a contract has recently been awarded by the Australian government. Although it is far too early to tell how this critical element will operate, it is encouraging that it emanates from an organisation that has grown up in a military environment in which there is very little scope for error, let alone failure. This is the Israeli organisation Elbit Industries, which has developed a battle-tested network system which may indeed provide the Australian Defence Force with the network capabilities sought.

The proof of success will lie in the front-line soldier receiving information that is valuable in a supportive sense, that is delivered in a form that is easily understood, and that is not obscured by an abundance of less relevant (even if technically interesting) data. These are things that make the soldier's task easier and less a matter of inspired guesswork than the provision of useful, advance information about 'what lies on the other side of the hill'.

As demonstrated by the long-lived AN/PRC-77 VHF FM radio, a reliable and simple-to-operate device is far more useful in the conventional battle situation than a radio with outstanding features that is complex to set up and operate. Additionally, the famous Handie-Talkie of World War Two, the BC-611, despite the limitations of a single crystal-controlled operating frequency and operation on the HF band, still performed a valuable service over short ranges.

Finally, while there may be many in the military arena who share the attitude of Henry Ford that 'history is more or less bunk', for this author, the position of George Santayana is a good deal more rational: 'Those who cannot remember the past are condemned to repeat it'. The repetition of analogous situations in military history does tend to suggest that there is much to be gained from careful reading of past events. The withdrawal from Gallipoli as a model for the successful withdrawal from Arnhem is just one example. Major-General Urquhart's study of the historical precedent of Gallipoli provided an invaluable insight into the solution of a very difficult military problem: extricating a large body of infantry from an untenable situation on the north bank of the Lower Rhine.

Postscript

Reflecting on the review undertaken in this work has highlighted some disquieting realities of warfare and the changing technology of communications that it has demanded. In particular, two communications devices that were encountered along the way, the first in Part 1, the BF trench transmitter and receiver, and the second at the end of Part 4, the SELEX/Marconi PRR, reveal some underlying functional and operational similarities. These are characteristics that have responded to the inherent symmetry of military demands for a simple, personal and private means of achieving conversations between fighters on the front line.

Clearly what has changed over the years is the technology, which has enabled a remarkable reduction in size and weight of apparatus. This has been achieved while operating range, simplicity of operation and reliability have all increased. What has not changed significantly is the manner in which radios are used to convey orders and receive reports of the disposition of the opposition. What also has not changed since the first human took up a spear to impale an opponent is that projectiles, however propelled, lead to a messy and almost inevitable final departure. In any such confrontation, the capacity to call on an ally to provide support, either by shouting or by radio, remains a fundamental component of the military ethos. Moreover, whether the source of the projectile be a Brown Bess or an electronically controlled deliverer of explosive ballistic armaments, the end point is the same—the incapacitation or death of an opponent while ensuring one's own survival.

The last 100 years of warfare and communication have made it brutally clear that there is very little difference between the hand-to-hand combat of apelike ancestors on the African grasslands and the most modern warfare with stand-off missiles or roadside IEDs. What is different is the technology used to assist in the military task, and perhaps its civilian and peaceful applications will in the end ensure that military action becomes less and less necessary or frequent. In saying that, perhaps the application of military network systems to the Internet and the use of military-developed digital devices for non-military applications will make the century ahead less likely to end in the Armageddon of prophecy. Regrettably, human character must make one somewhat pessimistic, and in this the famous French saying remains disturbingly true:

> *'Plus ça change, plus c'est la même chose.'*
>
> 'The more things change, the more things stay the same.'

Bibliography

Aitken, HGJ. *Syntony and Spark.* Wiley, New York, 1976.
Aitken, HGJ. *The Continuous Wave.* Princeton University Press, Princeton NJ, 1985.
Anglo-Boer War Museum. *The Second Boer War.* Anglo-Boer War Museum, Bloemfontein, South Africa: http://www.anglo-boer.co.za/intro/peace-treaty-vereeniging.php
Anno, SE & Einspahr, WE. *Command and Control: Lessons Learned: Etc* Air War College Research Report. http://www.dtic.mil/cgi-bin/GetTRDoc?AD=ADA202091
Arnold-Forster, M. *The World at War.* William Collins, London, 1973.
Austin, B. *Wireless in the Boer War.* Institution of Electrical Engineers, London, 1994.
Austin, B. 'HF propagation and clandestine communications during the Second World War', *Radio Bygones*, No 120, August-September 2009.
Australian Army. *The Division in Battle Pamphlet Number 7 Signals.* 1970. www.qsl.net/vk2dym/radio/Sigs70a.htm
Australian Army. *Adaptive Army.* Public Information Paper, Canberra, 2008.
Australian Army. *Signal Training, Vol. III, Pamphlet No. 5, Wireless Set No. 108 Mk I and II.* Melbourne, 1941.
Australian Corps of Signals. *Signals: Story of the Australian Corps of Signals.* Halstead Press, Sydney, 1944.
Australian Government. *Australia in the Vietnam War.* Commemorative website. http://vietnam-war.commemoration.gov.au/
(Author not listed). *Bowman – Personal Role Radio AN/PRC 343.* http://www.prc68/l/Bowman.shtml
(Author not listed). *Admiral Bezobrasov Strikes.* Russo-Japanese War Research Society. http://www.russojapanesewar.com/bezo-strikes.html
(Author not listed). *Printed Circuit Boards: How Products are Made.* http://www.madehow.com/Volume-2/Printed-Circuit-Board.html
(Author not listed). *Boat Anchor Manual Archive* 2012. http://bama.edebris.com/manuals/
(Author not listed). 'The American Institute of Electrical Engineers Annual Dinner and Mr Marconi', *The Electrical World and Engineer*, New York, Jan 18, 1902.
(Author not listed). 'Ernst Alexanderson', *Electronics Australia Yearbook 1975/76*, p. 14.
(Author not listed). 'Portable wireless telegraphy', *The Marconigraph*, Aug 1912, p. 185.
(Author not listed). *Year-Book of Wireless Telegraphy 1914.* Marconi Publishing Corporation, New York, 1914.
(Author not listed). 'Land 125 phase', *SoldierMod*, Vol. 4, 2010: www.soldiermod.com
(Author not listed). 'Thomas Alva Edison', *Tele-Technician*, Nov-Dec 1950.
(Author not listed). Raytheon, DMO sign JP2072 EPLRS Contract. *Australian Defence Monthly* 2011: http://www.australiandefence.com.au/archive/news-review-raytheon-dmo-sign-jp2072-eplrs-contract-adm-mar-2011

(Author not listed). Battlespace Communications System (Land) JP 2072 Phase 1: http://www.defence.gov.au/dmo/esd/jp2072/index.cfm?media=print

(Author not listed). *History of the Tank–World War 1–Interwar–World War 2*. Global Security: http://www.globalsecurity.org/military/systems/ground/tank-history1.htm

(Authors not listed). *Europe After World War 1: November 1918 to August 1931*. Legacy Publishers: http://history.howstuffworks.com/world-war-i/europe-after-world-war-1.htm

(Author not listed). 'Transistor pocket radio', *Radio Television Electronics*, January 1955.

(Author not listed). *History of Mobile/Cell Phone*. Radio-Electronics.com: http://www.radio-electronics.com/info/cellulartelecomms/history/mobile-cell-phone.php

(Author not listed). *History of the Royal Tank Regiment*. Royal Tank Regiment Association: http://www.royaltankregiment.com/en-GB/ww1.aspx

(Author not listed). *Air Force Network Integration History*. US Air Force: http://www.afnic.af.mil/library/factsheets/factsheet.asp?id=6847

(Author not listed). *Internet Access Guide*. Conniq.com: http://www.conniq.com/InternetAccess_Introduction.htm

AWA. *Military Set Type 101 Booklet 1063*. Amalgamated Wireless (Australasia) Ltd, Sydney, nd.

AWA. *Radio Set PRC-F1 Provisional User Handbook*. 1968. http://www.tuberadio.com/robinson/Manuals/PRC-F1_Handbook_Provisional.pdf

Baker, DC. 'Wireless telegraphy during the Anglo-Boer War of 1899–1902', *South African Military History Society Journal*, Vol 11, No. 2. http://samilitaryhistory.org/vol112db.html

Baker, WJ. *A History of the Marconi Company*. Methuen, London, 1970.

Barker, T. *Signals: History of the Royal Australian Corps of Signals 1788–1947*. Corps Committee, Royal Australian Engineers, Canberra, 1987.

Barnes, K. 'The Defence Signals Directorate: its roles and functions', *Australian Defence Force Journal*, No. 108, September-October 1994.

Barty-King, H. *Girdle Round the Earth*. Heinemann, London, 1979.

Bateman, J. *The History of the Telephone in NSW*. Telephone Collectors Society of NSW, 1980.

Bauer, AO. 'Receiver and transmitter developments in Germany: 1920 to 1945', *International Conference Publication No. 411*, Institution of Electrical Engineers, London, 1995.

Bauer, AO. *Some Aspects of the German Military 'Abwehr' Wireless Service*. Centre for German Communications and Related Technology 1920–1945. http://www.cdvandt.org/German%20Abwehr.pdf

Bean, CE. *Anzac to Amiens*. Penguin Books, Melbourne, 1993.

Beaumont, J. *Australia's War—1914/1918*. Allen & Unwin, Sydney, 1995.

Beesley, P. *Very Special Intelligence*. Sphere, London, 1977.

Beier, KP. *Virtual Reality: A Short Introduction* 2001. http://www-vrl.umich.edu/intro/

Bell, ATJ. 'The Battle for Crete: the tragic truth', *Australian Defence Force Journal*, Jul/Aug 1991.

Belrose, JS. 'Fessenden and Marconi: Their differing technologies and trans-Atlantic experiments during the first decade of this century', *IEE Conference: 100 Years of Radio*, Sept 1995, p. 32.

Bennet, G. *Naval Battles of the First World War* [1968]. Penguin Books, London, 2001.

Best, SP. *The Venlo Incident*. Frontline Books, London, 2009.

Billings, A. Transcript of Address to Watsonia School of Signals Museum, November 1979. Museum Archives.

Billings, A. Autobiography of Bert Billings, written 1974. Australian War Memorial File 419/10/6 Record 3DRL 6060.

Blaine, J. *An Introduction to Software Defined Radio*, 2009: http://www.ac0c.com/attachments/An_Introduction_to_Software_Defined_Radios_v2a.pdf

Blake, GG. *History of Radio Telegraphy and Telephony*. Radio Press, London, 1926.

Blainey, G. *The Tyranny of Distance*. Macmillan, Melbourne, 1967.

Blaxland, J. *Signals—Swift and Sure*. Royal Australian Corps of Signals Committee, 1998.

Blaxland, J. 'The Role of Signals Intelligence in Australian Military Operations 1939–1972', *Australian*

Army Journal, Vol. 11, No. 2, p. 203.
Bleakley, J. *The Eavesdroppers*. Australian Government Publishing, Canberra, 1991.
Bowyer, C. *Fighter Command*. Sphere, London, 1980.
BBC. *Twenty-Five Years of British Broadcasting 1922–1947*. BBC Publishing, London, 1947.
Bradley, MP. *Vietnam at War*. Oxford University Press, Oxford, 2009.
Brooks, K. 'Special Operations Executive: the B2 Spy Set', *VMARS Newsletter*, Issue 41, June 2005.
Brown, AC. *Bodyguard of Lies*. WH Allen, London, 1976.
Bryant, A. *Turn of the Tide, 1939–1943*. Fontana, London, 1957.
Buchan, J. *The Battle of the Somme: First Phase*. Thomas Nelson, London, 1916.
Bull, K. 'Samuel Morse', *Amateur Radio Action*, Vol. 11, No. 10, p. 45.
Bull, K. 'The father of modern radio', *Amateur Radio Action*, Vol 10, No. 4, p. 21.
Bull, K. 'Edison, Bell and the carbon microphone', *Amateur Radio Action*, Vol 8, No. 5, p. 43.
Burke, K. *With Horse and Morse in Mesopotamia*. A. & N.Z. Wireless Signal Squadron History Committee, Sydney, 1927.
Burton, N. 'Radio genesis Australia', *The Mariner*, March-April 1968, p. 206.
Bushby, R. *The Australian Army and the Vietnam War 1962–1972*. http://www.army.gov.au/ahu/docs/the_australian_army_and_the_vietnam_war_bushby.pdf
Cairncross, F. *The Death of Distance*. Orion Books, London, 1997.
Calvocoressi, A. *Top Secret Ultra*. Sphere, London, 1979.
Cameron, AR. The Story of the Overland Telegraph Line. Lecture to the South Australian Postal Institute, 11 October 1932, unpublished.
Campbell-Kelly, M & Aspray, W. *Computer: A History of the Information Machine*. Basic Books, New York, 1996.
Cawthorne, N. *Vietnam: A War Lost and Won*. Hinkler Books, Heatherton, VIC, 2003.
Cecil, M. 'Remember when ... we built wireless sets?' *Department of Defence Materiel 2006*. http://www.defence.gov.au/dmo/news/ontarget/apr06/rw.cfm
Chambers, EW. Letter dated 4 May 1972 and paper entitled 'Saga of Australian Wireless 1888 to 1920'. Unpublished correspondence, Marconi Archives, Great Badow, near Chelmsford, UK.
Champness, R. 'Behind the lines: a short history of spy radios in WWII', *Silicon Chip*, September-October 1998.
Chapman, EH. *Wireless Today*. Oxford University Press, Oxford, 1941.
Churchill, WS. *The World Crisis 1911 to 1918*. Four Square, London, 1923.
Churchill, WS. *The Second World War*, Vols 1–6, Cassell, London, 1948.
Clarke, AC. *Profiles of the Future*. Pan Books, London, 1973.
Claricoats, J. *World at their Fingertips*. Radio Society of Great Britain, 1967.
Clark, A. *Barbarossa*. Penguin Books, London, 1965.
Clarke, AC. *How the World was One*. Victor Gollancz, London, 1992.
Clarke, R. *A Brief History of the Internet in Australia* 2002. http://www.anu.edu.au/people/Roger.Clarke/II/OzIHist.html
Clarke, R. *Origins and Nature of the Internet in Australia*. Xamax Consultancy, 2004. http://www.anu.edu.au/people/Roger.Clarke/II/OzI04.html
Cleary, P. *The Men Who Came Out of the Ground*. Hachette, Sydney, 2000.
Collier, R. *The Sands of Dunkirk*. Fontana, London, 1961.
Collins, AF. *Radio Amateur's Handbook*. Thomas Y Crowell, New York, 1922.
Commonwealth Government. 'The Landing at Gallipoli: Report from Government Press Representative,' *Commonwealth of Australia Gazette*, No. 39, May 1915.
Connell, B. *The Return of the Tiger*. Pan, London, 1960.
Constable, A. *Early Wireless*. MIDAS, Tunbridge Wells, UK, 1980.
Corcoran, A.P. 'Wireless in the trenches', *Popular Science Monthly*, May 1917.
Cornwell, J. *Hitler's Scientists*. Penguin Books, London, 2004.
Coulthard-Clark, CD. 'Australia's war-time security service', *Australian Defence Force Journal*, No. 16,

May-Jun 1979.
Cox, J. Raytheon Creates First Mobile Ad Hoc Battlefield Net etc. *Network World* 2009. http://www.networkworld.com/community/node/39588
Cross, J. *Red Jungle*. Robert Hale, London, 1958.
Crowther, JG & Whiddington, R. *Science at War*. HMSO, London, 1947.
Davies, MJ. 'Military intelligence at Gallipoli: "a leap in the dark"', *Australian Defence Force Journal*, No. 92, Jan-Feb 1992.
Deacon, R. *A History of the British Secret Service*. Granada, London, 1969.
Deighton, L. *Blitzkrieg: From the rise of Hitler to the fall of Dunkirk*. Triad Grafton, London, 1979.
DeLacy, B. 'Who did invent the superheterodyne?', *HRSA Radio Waves*, April 1995, p. 25.
Department of Defence. *Force Structure Review* 1991: http://www.defence.gov.au/oscdf/se/publications/Force%20Structure%20Review%201991_opt.pdf
Department of the Army. *Radio Set AN/PRC 64A Technical Manual* TM 11-5820-552-15. Washington, DC, 1970.
DeSoto, CB. *200 Meters and Down: The Story of Amateur Radio*. Amateur Radio Relay League, Hartford, CT, 1936.
DeVries, L. *The Book of Telecommunications*. John Murray, London, 1962.
Douglas, A. 'Who invented the superheterodyne?', *Proceedings of the Radio Club of America*, Vol. 64, No. 3, Nov 1990.
Duncan, R & Drew, CE. *Radio Telegraphy and Telephony*. John Wiley & Sons, New York, 1929.
Dunn, P. *Central Bureau in Australia During WW2*. Australia at War: http://www.ozatwar.com/sigint/cbi.htm
Dunn, P. *Coast Watch Organisation or Combined Field Intelligence Service Section 'C' of the Allied Intelligence Bureau*. Australia at War: http://www.ozatwar.com/sigint/coastwatchers.htm
Durrant, L. *The Seawatchers*. Angus & Robertson, Sydney, 1986.
Eaton, JP & Haas, CA. *Titanic: Destination Disaster—The Legends and the Reality*. Patrick Stephens, Wallingborough, UK, 1987.
Edwards, JR & Argus, JA (eds). *Australian Telecommunications at War: Special Issue, Radio & Electrical Retailer*, Mingay Publishing, Sydney, 1946.
Einstein, A & Infield, E. *The Evolution of Physics*. Cambridge University Press, Cambridge, 1938.
Eisenhower, DD. *Crusade in Europe*. William Heinemann, London, 1948.
Enever, T. *Britain's Best Kept Secret: Ultra's Base at Bletchley Park*. Alan Sutton, London, 1994.
Edwards, R. *Panzer*. Arms and Armour Press, London, 1993.
Erb, E. *Radiomuseum Online*. Megen, Switzerland: http://www.radiomuseum.org/
Erskine-Murray, J. *Wireless Telegraphy*. Crosby Lockwood, London, 1914.
Eunson, M. 'The early telegraph systems', *Amateur Radio Action*, Vol. 9, No. 7, p. 18.
Eunson, M. 'Telegraphy, Morse code and all that', *Amateur Radio Action*, Vol. 9, No. 9, p. 27.
Evans, M & Parkin, R. *Future Arms, Future Challenges*. Allen & Unwin, Sydney, 2004.
Evans, ND. *Military Gadgets: How Advanced Technology is Transforming Today's Battlefield and Tomorrow's*. Prentice Hall, New York, 2004.
Feldt, E. *The Coast Watchers*. Oxford University Press, Melbourne, 1946.
Fleming, JA. *Radio Telegraphy and Radio Telephony*. Longmans Green, London, 1916.
Fleming, JA. *The Thermionic Valve*. Wireless Press, London, 1919.
Fleming, JA. *Principles of Electric Wave Telegraphy and Telephony*. Longmans Green, London, 1919.
Fong, K. 'Portable battlefield communications', *Australian Defence Journal*, Nov. 2008: http://www.army.gov.au/ahu/docs/the_australian_army_and_the_vietnam_war_bushby.pdf
Foot, MRD. *SOE: The Special Operations Executive*. BBC Publishing, London, 1984.
Forbes, C. *The Korean War: Australia in the Giants' Playground*. Pan Macmillan, Sydney, 2010.
Fordred, LL. 'Wireless in the Second Anglo-Boer War 1899–1902', *Trans SA Institute of Electrical Engineers*, September, 1997.
Forty, G. *The British Tank: A Photographic History 1916–1986*. Tank Museum, Birlings, Kent, UK, 1986.

Bibliography

Fox, B & Webb, J. 'Colossal sdventures', *New Scientist*, Vol. 154, No. 2081, 1997.
Frater, M & Ryan, M. *Communications: Electronic Warfare and the Digitised Battlefield*. Working Paper No. 116, 2001: www.army.gov.au/lwsc/Docs/WP_116.pdf
Gallaway, J. *The Last Call of the Bugle: The Long Road to Kapyong*. University of Queensland Press, Brisbane, 1994.
Garratt, GRM. *The Early History of Radio: From Faraday to Marconi*. Institution of Electrical Engineers, London, 1994.
Garcia, C et al. *Tele-immersion: the Future of Internet Tele-communications*, 2002: www.dcs.napier.ac.uk/~mm/socbytes/feb2002_i/5.html
Garlinski, J. *Intercept: The Enigma War*. Dent, London, 1979.
Gates, W, Myhrvold, N & Rinearson, P. *The Road Ahead*. Viking, New York, 1995.
Geeves, P. 'Australia's radio pioneers' (in four parts), *Electronics Australia*, May 1974, p. 26; June 1974, p. 30; July 1974, p. 34; August 1974, p. 50.
Geeves, P. 'Marconi and Australia', *AWA Technical Review*, Vol. 15, No. 4, 1974, p. 131.
Gernsback, S. *S. Gernsback's Radio Encyclopaedia 1927*. Vintage Radio, Palos Verdes, CA, 1974.
Giles, JM. *George Augustine Taylor: Some chapters in the life*, Supplement to *Building: Lighting: Engineering*, 24 November 1957.
Gill, EW. *War, Wireless and Wangles*, Basil Blackwell, London, 1934.
Gobert, W. 'The evolution of service strategic intelligence 1901 to 1941', *Australian Defence Force Journal*, No. 92, January-February 1992.
Grandin, R. *The Battle of Long Tan*. Allen & Unwin, Sydney, 2004.
Gray, ED. 'The story of tanks: De Mole's travelling caterpillar fort', *Argus*, 9 August 1922.
Greenacre, JW. *Assessing the Reasons for Failure: 1st British Airborne Division Signals Communications during Operation 'Market Garden'*: http://www.tandfonline.com/doi/pdf/10.1080/1470243042000344777
Grey, J. *The Australian Army*. Oxford University Press, Melbourne, 2001.
Grey, J. *A Military History of Australia*, 3rd edn. Cambridge University Press, Cambridge, 2008.
Hafner, K & Lyon, M. *Where Wizards Stay Up Late: The Origins of the Internet*. Simon & Schuster, New York, 1996.
Ham, R. 'Valve and vintage: WS No. 19', *Practical Wireless*, February 1994.
Ham, R. 'Valve and vintage: the B2 transceiver', *Practical Wireless*, July 1995.
Ham, P. *Vietnam: The Australian War*. Harper Collins, Sydney, 2007.
Hammond, J. 'The father of FM', *73 Magazine*, Feb. 1982, p. 50.
Hancock, HE. *Wireless at Sea*. Marconi International Marine Communications Company Ltd, Chelmsford, UK, 1950.
Hankins, R. *The Wireless Set No. 31—Boring?*: http://www-users.kawo2.rwth-aachen.de/~banish/website/article_ws_31.html
Harclerode, P. *Para! Fifty Years of the Parachute Regiment*. Orion, London, 1992.
Harcourt, E. *Taming the Tyrant: The First One Hundred Years of Australia's International Communication Services*. Allen & Unwin, Sydney, 1987.
Hare, D. *Pronto in South Vietnam 1962–1972*. Digital book from author: http://pronto.au104.org/
Harfield, A. *Pigeon to Packhorse*. Picton Publishing, Chippenham, UK, 1989.
Harris Corporation. *AN/PRC-150 – Type-1 HF Radio*: http://rf.harris.com/capabilities/tactical-radios-networking/anprc-150.asp
Harris Corporation. *AN/PRC-152 – Type-2 VHF Radio*: http://rf.harris.com/capabilities/tactical-radios-networking/an-prc-152/
Haugland, FK. *The Wireless Services in the Grouse Group etc*. Debriefing Report 1943 from National Archives at Richmond, UK.
Haukelid, K. *Skis Against the Atom*. William Kimber, London, 1954.
Hawker, P. 'The secret of wartime radio', *Amateur Radio*, March 1983, p. 11.
Hawker, P. 'Clandestine radio: the early years', *Wireless World*, February 1982.

Hawkhead, JC. *Handbook of Technical Instruction for Wireless Telegraphists*. Marconi Press Agency, London, 1913.
Hawkhead, JC & Dowsett, HM. *Handbook of Technical Instruction for Wireless Telegraphists*. Wireless Press, London, 1915.
Hawkhead, JC & Dowsett, HM. *Handbook of Technical Instruction for Wireless Telegraphists*, 7th edn. Iliffe, London, 1943
Hawkhead, JC & Dowsett, HM. *Handbook of Technical Instruction for Wireless Telegraphists*, 9th edn. Iliffe, London, 1950.
Hill, J. *Wireless in the Trenches*. Historic Radio Society of Australia, nd.
Hill, J. *The Cat's Whisker: 50 Years of Wireless Design*. Oresko, London, 1978.
Hill, J. *Radio! Radio!* Sunrise Press, London, 1993.
Hinsley, FH & Stripp, A (eds). *Code Breakers: The Inside Story of Bletchley Park*. Oxford University Press, New York, 1994.
Hinsley, FH. & Thomas, EE. *British Intelligence in the Second World War*. Cambridge University Press, Cambridge, 1988.
HM Signal School. *Admiralty Handbook of Wireless Telegraphy*. HMSO, London, 1925.
HM Signal School. *Admiralty Handbook of Wireless Telegraphy*, Vols 1 & 2. HMSO, London, 1938.
Holst, A. 'Wireless in the 1914/18 War', *Amateur Radio*, April 1990.
Horner, D. *From Korea to Pentropic: the Army in the 1950s and early 1960s*: http://www.army.gov.au/ahu/docs/The_Second_Fifty_Years_Horner.pdf
Horrocks, Sir B. *A Full Life*. Fontana, London, 1960.
Hough, R & Richards, D. *The Battle of Britain: The Jubilee History*. John Curtis, London, 1989.
Howard, WL. *Spy Radios of WWII*: http://www.wlhoward.com/
Howard, C. 'Warfighters on the Digital Battlefield require etc.' *Military & Aerospace Electronics*, June 2010: http://integrator.hanscom.af.mil/2010/June/06172010/06172010-17.htm
Hoy, HC. *40 OB or How the War was Won*. Hutchinson, London, 1934.
Hoyt, EP. *Japan's War: The Great Pacific Conflict*. Arrow, London, 1986.
Isted, GA. 'Guglielmo Marconi and the history of radio' (2 parts), *GEC Review*, Vol. 7, No. 1, 1991; Vol. 7, No. 2, 1991.
Ingersoll, R. *Top Secret*. Partridge, London, 1946.
Jensen, PR. *In the Footsteps of Marconi: Early Radio*. Kangaroo Press, Sydney, 1994.
Jensen, PR. *From the Wireless to the Web*. UNSW Press, Sydney, 2000.
Jensen, PR. 'Clandestine communications and the Paraset', *Radio Bygones*, No. 118, April-May 2009.
Jensen, PR. 'Clandestine communications and a collection in Kent', *Radio Bygones*, No. 125, June-July 2010.
Jolly, WP. *Marconi: A Biography*. Constable, London, 1972.
Jahnke, DA & Fay, KA (eds). *From Spark to Space: A Pictorial Journey Through 75 Years of Amateur Radio*. American Radio Relay League, Newington, CT, 1989.
James, RR. *Chindit: The Explosive Truth About the Last Wingate Expedition*. Sphere, London, 1981.
Johnson, B. *The Secret War*. Arrow, London, 1978.
Jones, BE. *Small Electric Apparatus* Cassell & Co. Ltd, London, 1913.
Jones, RV. *Most Secret War*. Hamish Hamilton, London, 1978.
Jung, K. *Personal On-line Radio Museum*. Saulheim Germany: http://www.kpjung.de/e_ego.htm
Juniper, D. 'The First World War and radio development', *History Today*, May 2004.
Kahn, D. *The Codebreakers: The Story of Secret Writing*. Macmillan, New York, 1967.
Kahn, D. *Hitler's Spies: German Military Intelligence in World War II*. Arrow, London, 1978.
Kates, J & Smith, N. 'Thomas Edison, radio prophet', *Amateur Radio Action*, Vol. 4, No. 9, p. 18.
Keller, J. 'Military crypto-modernisation leads to applications etc', *Military and Aerospace Electronics*: http://www.militaryaerospace.com/articles/2011/11/military-crypto-modernization.html
Kingsford-Smith, Sir Charles & Miller, HC. *The Southern Cross Story/Early Birds* [1932]. Seal Books,

Bibliography

Sydney, 1995.
Knagge, G. Digital Subscriber Loop DSL and ADSL website: http://www.geoffknagge.com/uni/elec351/351assign.shtml
Knight, L. 'The coming of the superhet' (2 parts), *Radio Bygones*, Christmas 1990, p. 10; Feb-March 1991, p. 28.
Kozacuk, W. *Enigma: How the German Cipher Machine was Broken and How it was Read by the Allies in World War Two*. Arms and Armour, London, 1984.
Ladd, J & Melton, K. *Clandestine Warfare: Weapons and Equipment of the SOE and OSS*. Guild, London, 1988.
Laffin, J. *Special and Secret*.Time-Life Books, Sydney, 1990.
Lawson, RS. '1899–1902: The Boer War', *Australian Defence Journal*, No. 12, Sept-Oct 1978.
Leasor, J. *Green Beach*. Corgi, London, 1975.
Lee, B. 'Radio intelligence developments during World War One and between the wars', *Antique Radio Resource*: http://antiqueradios.com/chrs/journal/intelligence.html
Leslie, M. RAVEN: Historical Notes. Personal communication, February 2012.
Lewin, R. *Ultra Goes to War*. Grafton, London, 1988.
Licklider, JCR. 'Man–computer symbiosis', *IRE Transactions on Human Factors in Electronics*, Vol. HFE-1, March 1960: http://groups.csail.mit.edu/medg/people/psz/Licklider.html
Liddell Hart, BH. *History of the First World War* [1930]. Pan, London, 1973
Liddell Hart, BH. *History of the Second World War* [1970]. Pan, London, 1973.
Linton, J. 'The last wireless ANZAC', *Amateur Radio*, April 1990.
Lorain, P. *Secret Warfare: The Arms and Techniques of the Resistance*. Orbis, Lonon, 1983.
Lord, W. *Lonely Vigil: Coastwatchers of the Solomons*. Pocket Books, New York, 1977.
MacKeand, JCB & Cross, MA. 'Wide-band high-frequency signals from Poldhu', *IEE Conference: 100 Years of Radio*, Sept 1995, p. 26.
McKie, R. *The Heroes*. Angus & Robertson, Sydney, 1967.
Mackenzie, C. *Gallipoli Memories* [1929]. Panther, London, 1965.
Mackenzie, W. *Secret History of SOE: Special Operations Executive 1940–1945*. St Ermin's Press London, 2000.
MacKinnon, C. *The 3BZ Coast Watchers' Wireless Set*: http://www.qsl.net/vk2dym/radio/3BZa.htm
MacKinnon, C. *Wireless Set No. 108 Marks I, II and III*: http://www.qsl.net/vk2dym/radio/NO.108.htm
Macksey, KJ. *Panzer Division: The Mailed Fist*. Macdonald, London, 1968.
McKay, F. *Traeger: The Pedal Radio Man*. Boolarong Press, Moorooka, QLD, 1995.
McKernan, M & Brown, M. *Two Centuries of War and Peace*. Australian War Memorial, Canberra, 1988.
McMahon, M. *Vintage Radio: A Pictorial History of Wireless and Radio 1887–1929*. Vintage Radio, Palos Verdes Peninsula, CA, 1973.
Marceil, WS. 'The first radio broadcast', *Radio Bygones*, Christmas 1992, p. 4.
Marconi, G. *The Progress of Electric Space Telegraphy*. Royal Institution, London, 1902.
Marcus, G. *The Maiden Voyage: A Complete and Documented Account of the TITANIC Disaster*. George Allen& Unwin, London, 1969.
Marples, G. *The History of DSL Internet Access: A Race for Technological Speed*: http://www.thehistoryof.net/history-of-dsl.html
Masefield, J. *Gallipoli*. Heinemann, London, 1917.
Masterman, JC. *The Double Cross System 1939–1945*. Granada, London, 1979.
Matthews, W. 'Teaching an old radio new tricks', *Defence News* 2009: http://www.defensenews.com/story.php?i=3979778
Metcher, O. 'The Landing on Gallipoli 62 Years Ago', based on extracts from diary held at Watsonia School of Signals Museum Archives, VIC.
Meulstee, L. 'Historical development of British Army man-pack radio sets' (2 parts), *HRSA Newsletter*,

July 1987; October 1987.

Meulstee, L. *Wireless for the Warrior*, Vols 1-4. Wimborne Publishing, Wimborne, UK, 2004.

Meulstee, L. *Wireless for the Warrior Compendium 1*. EMAUS, Groenlo, Netherlands, 2009.

Meulstee, L. 'Larkspur'. *Wireless for the Warrior*: http://wftw.nl/larkspur/larkspur1.html

Miles, WGH. *Admiralty Handbook of Wireless Telegraphy 1925*. HMSO, London, 1925.

Miller, CE. 'Radio goes to war', *Radio Bygones*, Aug-Sept 1989, p. 3.

Miller, J. *The Shape of Bits to Come*. Spectrum Online: http://sss-mag.com/G3RUH/index1.html

Ministry of Information. *By Air to Battle: The Official Account of the British First and Sixth Airborne Divisions*. HMSO, London, 1945.

Moorehead, A. *Gallipoli*. Wordsworth Military Library series, Wordsworth, Ware, UK, 1997.

Moyle, A. *Clear Across Australia: A History of Telecommunications*. Thomas Nelson, Melbourne, 1984.

Murdoch, KA. Letter to the Prime Minister of the Australian Commonwealth. Committee of Imperial Defence, September 1915, National Archive, Richmond, UK.

Murray, J. *Calling the World: The First 100 years of Alcatel Australia 1895-1995*. Focus, Double Bay, NSW, 1995.

Muscio, WT. *Australian Radio: The Technical Story 1923-1983*. Kangaroo Press, Sydney, 1984.

Nalder, RFH & Royal Signals Institution. *The Royal Corps of Signals: A History of its Antecedents and Development (circa 1800-1955)*. Royal Signals Institution, London, 1958.

National Security Agency (NSA). *The Origination and Evolution of Radio Traffic Analysis: the World War 1 Era*: http://www.nsa.gov/public_info/_files/cryptologic_quarterly/trafficanalysis.pdf

Neyland, B. 'Memoirs & diaries: a wireless operator', in CB Purdom (ed.), *Everyman at War: Sixty Personal Narratives of the War*. JM Dent, London, 1930.

Occleshaw, M. *Armour Against Fate: British Military Intelligence in the First World War*. Columbus Books, London, 1989.

Odgers, G. *100 Years of Australians at War*. [Endorsed by the Australian Defence Force.] Lansdowne, Sydney, 1999.

Odgers, G. *Remembering Korea: Australians in the War of 1950-1953*. Lansdowne, Sydney, 2000.

O'Neill, R. The Korean War and Australia. Australian War Memorial Oration, 2003.

Ostrovsky, YI. *Holography and its Applications*. Mir, Moscow, 1977.

O'Toole, I. *Reference Notes: Various Military Radios*. Castle Hill Military Radio Museum, Castle Hill (Sydney): http://www.vk2bv.org/museum/

Pakenham, T. *The Boer War*. Weidenfeld & Nicolson, London, 1979.

Parker, GE. *Report: Sparrow Force East Timor March to April 1942*. From Royal Australian Signals Museum Archives 2011.

Patriot Files Website: *Electronic Warfare in WW1*: http://www.patriotfiles.com/forum/showthread.php?t=110034

Payne, D & Reeve, J. *Armoured Cars and Tanks on the Western Front in the Great War*: http://www.westernfrontassociation.com/great-war-on-land/73-weapons-equipment-uniforms/78-armoured-cars-western.html

Pelkey, J. *Entrepreneurial Capitalism and Innovation: A History of Computer Communications 1968-1988*: http://www.historyofcomputercommunications.info/Book/BookIndex.html

Perquin, J-L. *The Clandestine Radio Operators: SOE, BCRA, OSS*. Histoire & Collections, Paris, 2011.

Perrett, B. *Lightning War: A History of Blitzkrieg*. Jove, New York, 1989.

Peverett, AM. 'Some early radio receivers', *Wireless World*, Nov 1972, p. 510.

Pickworth, G. 'The spark that gave radio to the world', *Electronics and Wireless World*, November 1993, p. 937.

Pickworth, G. 'Marconi's 200 kW trans-Atlantic transmitter', *Electronics and Wireless World*, January 1994, p. 28.

Pickworth, G. 'Germany's Imperial wireless system', *Electronics and Wireless World*, May 1993.

Pidgeon, G. *The Secret Wireless War: The Story of MI6 Communications*. Arundel Books, Richmond, UK, 2008.

Bibliography

Pierce, S.D. *Radio Set AN/PRC 64 Development and Test Program*. US Army Limited War Laboratory, Aberdeen, MD, USA, 1967: http://www.dtic.mil/cgi-bin/GetTRDoc?AD=AD813297&Location=U2&doc=GetTRDoc.pdf

Powell, G. *The Devil's Birthday: The Bridges to Arnhem, 1944*. Macmillan, London, 1984.

Poole, T. 'Wireless Set No. 109', *HRSA Newsletter*, July 1990.

Price, HE. *Digital Communications*: ftp://ftp.tapr.org/ss/QEX95.SS.pdf

Priestley, RE. *The Signal Service in the European War of 1914 to 1918*. Royal Engineers and Signals Association, Chatham, UK, 1921.

Prince, D. *Current Aussie Radio Equipment: circa 1966*, transferred by Dave Prince: http://www.aulro.com/afvb/990106-post31.html

Raggett, RJ (ed.). *Jane's Military Communications*. Editions from 1978 to 2011, Jane, London.

Raines, RR. *Getting the Message Through: A Branch History of the US Army Signals Corps*. Center of Military History, Washington, DC, 1996.

Raleigh, W. *The War in the Air Vol. 1*. Imperial War Museum, London: http://ia600309.us.archive.org/27/items/warinairbeingst00jonegoog/warinairbeingst00jonegoog.pdf

Ramsay Silver, L. *The Heroes of Rimau: Unravelling the Mystery of One of World War II's Most Daring Raids*. Sally Milner Publishing, Binda, NSW, 1990.

Ramsay Silver, L. *Krait: The Fishing Boat that Went to War*. Sally Milner Publishing, Binda, NSW, 1992.

Remarque, EM. *All Quiet on the Western Front* [1929]. Mayflower-Dell, London, 1963.

Reuvers, P & Simons, M. *Crypto Museum On-Line*: http://www.cryptomuseum.com/index.htm

Riordan, M & Hoddeson, L. *Crystal Fire: The Birth of the Information Age*. WW Norton, New York, 1997.

Robinson, CR. 'Memory of Crete', *Australian Defence Force Journal*, No. 88, Jul/Aug, 1991.

Robinson, R. Wireless Set No. 19 (Aust.). Information Note Wyong Field Day 2005.

Robinson, R. *A510 Trials*: http://www.shlrc.mq.edu.au/~robinson/Information/A510_Trials.html

Robinson, T. 'A contact too far? Practical trials of Arnhem battlefield communications,' *RadCom*, Feb 2005 RSGB: www.rsgb.org

Robison, SS. *Manual of Wireless Telegraphy (Radio) for the Use of Naval Electricians*. Lord Baltimore Press, Baltimore, MD, 1913.

Ross, J. 'EH Armstrong: FM pioneer', *The Broadcaster*, July 1986.

Ruby, M. *F Section SOE: The Buckmaster Networks*. Grafton, London, 1988.

Ryan, C. *A Bridge Too Far*. Coronet, London, 1974.

Ryan, M & Frater, M. *A Tactical Communications System for Future Land Warfare*. Working Paper No. 109, Land Warfare Studies Centre, 2000: http://www.army.gov.au/lwsc/docs/wp%20109.pdf

Sale, A. 'The Colossus of Bletchley Park', *IEE Review*, March 1995.

Sale, A. *The Lorenz Cipher and how Bletchley Park Broke It*: http://www.codesandciphers.org.uk/lorenz/fish.htm

Saunders, IL. 'The unidyne', *Radio Bygones*, Oct-Nov 1992, p. 21.

Schatzkin, P. *The Farnsworth Chronicles*: http://www.farnovision.com/chronicles/tfc-secret.html

Schonland, BFJ. 'W/T RE', *Wireless World*, No. 76, Vol. VII, July 1919.

Shawsmith, A. 'Edwin Howard Armstrong', *Amateur Radio Action*, Vol. 5, No. 2, p. 29.

Signal Service (France). 'W/T sets forward spark 20 watts B front and rear ("loop set")', *Technical Instructions*, No. 4, November 1917.

Signal Service (France). 'Tuner, Short Wave, Mark 3*', *Technical Instructions*, No. 13, December 1918.

Simons, RW. 'Guglielmo Marconi and early systems of wireless communications', *GEC Review*, Vol. 11, No. 1, 1996.

Smith, T. 'The origins of Morse', *Practical Wireless*, Feb 1986, p. 36.

Standage, T. *The Victorian Internet: The Remarkable Story of the Telegraph and the Nineteenth Century's On-Line Pioneers*. Weidenfeld & Nicholson, London, 1998.

Starks, D. *Additional Comments*: http://www.armyradio.com/arsc/customer/pages.php?pageurl=/publish/Articles/US_Military_Portable_Radios/Additional_Comments.htm

Starks, D. *AN/PRC 25—A Forgotten Legend*: http://www-users.kawo2.rwth-aachen.de/~banish/website/article_prc25.html

Sterling, GE. *The Radio Manual*. Van Nostrand, New York, 1928.

Sterrenburg, FAS. *The Oslo Report 1939—Nazi Secret Weapons Forfeited*: http://www.v2rocket.com/start/chapters/peene/oslo_report.html

Stewart, A. *Persian Expedition: The Australians in Dunsterforce, 1918*. Australian Military History Publications, Loftus, NSW, 2006.

Stokes, JW. *70 Years of Radio Tubes and Valves*. Vestal Press, New York, 1982.

Story, AT. *The Story of Wireless Telegraphy*. Hodder & Stoughton, London, 1904.

Sumner, D. 'Radio propagation at the Battle of Arnhem', *RadCom*, Nov. 2005 RSGB: www.rsgb.org

Sweeney, WM. *Wireless Telegraphy (for Professional or Amateur Students)*. EW Cole, Melbourne, 1920.

Swinson, A. *Defeat in Malaya: The Fall of Singapore*. Macdonald &Co., London, 1969.

Tasker, A. *US Military Portable Radios* 2000: http://www.nj7p.org/history/portable.html

Taylor, P. *An End to Silence: The Building of the Overland Telegraph Line from Adelaide to Darwin*. Methuen, Sydney, 1980.

Taylor, AJP *The First World War*. Penguin Books, Harmondsworth, UK, 1962.

Thyer, JH. *The Royal Australian Corps of Signals 1906 to 1918*. Self-published. Erindale, SA, 1974.

Thomson, A. *The Singing Line: Tracking the Australian Adventures of my Intrepid Great-Great-Grandmother*. Random House, London, 1999.

Tuohy, F. *The Secret Corps: A Tale of 'Intelligence' on All Fronts*. John Murray, London, 1920.

Urquhart, RE. *Arnhem*. Pan, London, 1958.

Vizard, F & Scott, P. *21st Century Soldier: The Weaponry, Gear, and Technology of the Military in the New Century*. Time Inc. Home Entertainment, New York, 2002.

Wain, N. 'Wireless signalling in the Australian Army' (10 parts), *Radio Waves HRSA*: Part 1, January 2001; Part 2, April 2001; Part 2B, July 2001; Part 3, October 2001; Part 4, January 2002; Part 5, April 2002; Part 6, July 2002; Part 7, January 2003; Part 8, October 2003; Part 9, July 2004; Part 10, January 2005.

Wakai, N. *Dawn of Radio Technology in Japan*. International Conference Publication No. 411, Institution of Electrical Engineers, London, 1995.

Wake, N. *The White Mouse*. Macmillan, London, 1985.

Walters, LC. 'Shannon, coding and spread spectrum', *Electronics and Wireless World*, March 1989.

Wander, T. 'Wireless takes to the road' (3 parts), *Radio Bygones*, Aug-Sept 1989, p. 20; Oct-Nov 1989, p. 27; Feb-March 1990, p. 16.

War Office, UK. *Signal Training*, Vol. 3 (Pamphlet No. 24: Aerials). The Army Council, 1939.

War Office, UK. Signal Training, Vol. 3 (Pamphlet No. 11: Wireless Telegraph Set 'C' Mark II). The Army Council, 1926.

Ward, HJB. 'Wireless waves in the world's war', in *Yearbook of Wireless Telegraphy and Telephony*, Wireless Press, London, 1916.

Wedgwood, A. 'Near vertical incidence skywave communications', *VMARS Newsletter*, Issue 16, April 2001

Wedlake, GEC. *SOS: The Story of Radio-Communication*. Wren, Melbourne, 1973.

Welchman, G. *The Hut Six Story: Breaking the Enigma Codes*. Penguin Books, Harmondsworth, UK, 1982.

Wells, HG. 'The Land Ironclads' [1903], in *Short Stories of HG Wells*. Penguin Books, Harmondsworth, UK, 1958.

West, N. *GCHQ: The Secret Wireless War 1900–86*. Coronet, Sevenoaks, UK, 1987.

White, H. 'Four decades of the defence of Australia: reflections on Australia's defence policy over the past 40 years', Chapter 11 in *History As Policy: Framing the Debate on the Future of Australia's*

Bibliography

Defence Policy, Proceedings of the Strategic and Defence Studies Centre's 40th Anniversary Seminar: http://epress.anu.edu.au/sdsc/hap/mobile_devices/index.html

Williams, HS. *Practical Radio*. Funk & Wagnalls, New York, 1924.

Williams, C. *Doctrine, Training and Combat with the 1st Battalion, The Royal Australian Regiment, 1965–1966*: http://www.army.gov.au/ahu/docs/The_Australian_Army_and_the_Vietnam_War_Williams.pdf

Williams, N. *Australia's Radio Pioneers*. Federal Press, Sydney, 1994.

Winch, D. Personal communication by email 02 December 2011, 20 December 2011.

Winter, D. *25 April 1915: The Inevitable Tragedy*. University of Queensland Press, Brisbane, 1994.

Winterbotham, FW. *The Ultra Secret*. Futura, London, 1974.

Wright, P. *Spycatcher*. Heinemann, Melbourne, 1987.

Wright, K. 'The design was not passed on', *Military History On-line*: http://www.militaryhistoryonline.com/wwi/articles/designnotpassedon.aspx

Wolff, L. *In Flanders Fields*. Longmans Green, London, 1959.

Yardley, HO. *The American Black Chamber*. Bobbs-Merrill, Indianapolis, IN, 1931.

Yardley, HO. *The Chinese Black Chamber*. Houghton Mifflin, Boston, MA, 1983.

Young, P (ed.). *The Decisive Battles of World War II*. Bison Books, London, 1989.

Glossary of Abbreviations

@	at	AP	armour-piercing
AA	anti-aircraft	APERS	anti-personnel
AB	air burst	APM	anti-personnel mine
ABC	atomic, biological, chemical (replaced by chemical, biological, radiological [CBR] and nuclear, biological, chemical [NBC])	APU	auxiliary power unit
		AR	assault rifle
		ARQ	automatic request for retransmission
		asap	as soon as possible
ABIT	advanced built-in test	ASCII	American Standard Code for Information Interchange
ABM	anti-ballistic missile		
AC	alternating current	AUG	Armee-Universal-Gewehr
ACARS	aircraft communicating and reporting system	AWACS	Airborne Warning and Control System
ACH	advanced combat helmet (MICH TC-2000 Combat Helmet)	AWOL	absent without official leave
		BAE	British Aerospace
		BAR	M1918 Browning Automatic Rifle
ACOG	advanced combat optical gunsight		
		BASE	British Aerospace Systems and Equipment
ACRV	armoured command and reconnaissance vehicle		
		BC	battery commander
ACS	artillery communications system	BFO	beat frequency oscillator
		BFV	Bradley Fighting Vehicle
ACV	armoured combat vehicle	BIFF	battlefield identification friend or foe
ADA	Air Defense Artillery (US)		
ADF	Australian Defence Force	BIOS	basic input output system
ADI	Australian Defence Industries	BITE	built-in test equipment
AEF	Allied Expeditionary Force	BMS	battlefield management system
AFC	Australian Flying Corps		
ah	ampere-hour	BPS	battery power source
AIF	Australian Imperial Force (Australia, WWI)	C3I	command, control, communications, and intelligence
ALE	automatic link establishment		
AM	amplitude modulation	CAD	computer-aided design
ANZAC	Australia and New Zealand Army Corps	CAS	close air support
		CBR	chemical, biological, radiological
ANZUS	Australia New Zealand United States (treaty)		
		CCD	charge-coupled device
		CCP	computer control panel

Glossary of Abbreviations

CDU	computer display unit	FCC	fire-control computer
CECOM	communications–electronics command	FFSS	future fighting soldier system
		FM	frequency modulation
CEU	computer electronics unit	FMJ	full metal jacket
CMOS	complementary metal oxide semiconductor	FO	forward observer
		FOB	forward operating base
CNVD	clip-on night vision device	FOO	forward observation officer
COB	close of business	FTP	file transfer protocol
CO	commanding officer	fubar	fouled up beyond all repair
COMZ	communications zone	GB	gigabyte
COS	chief of section	GHz	gigahertz
CofS	chief of staff	GMT	Greenwich Mean Time
COTS	commercial-off-the-shelf	GP	general purpose
Cpl	Corporal	GPMG	general purpose machine-gun
CPU	central processing unit	GPS	global positioning system
CRT	cathode-ray tube	H&K	Heckler & Koch
CS	communications subsystem	HALO	high altitude low opening
CVSD/CVSDM	continuously variable slope delta modulation	HE	high explosive
		HF	high frequency (3 to 30 MHz)
CW	continuous wave	HMG	heavy machine-gun
DA	double action	HP	high power; hollow point
DARPA	Defense Advanced Research Projects Agency (US)	HT	high tension
		HUD	head up display
DC	direct current	Hz	hertz
DDU	digital display unit	IC	integrated circuit
DHSS	data handling subsystem	ICBM	intercontinental ballistic missile
DMA	direct memory access		
DOS	disk operating system	IED	improvised explosive device
DSP	digital signal processing	IFF	identification friend or foe
DZ	drop zone	IR	infra-red
ECCM	electronic counter-counter measure(s)	IRBM	intermediate-range ballistic missile
ECM	electronic counter-measure(s)	IRQ	interrupt request
ELINT	electronic intelligence	JLTA	Javelin composite launch tube
EM	electromagnetic	JSSAP	Joint Services Small Arms Program (US)
EMI	electromagnetic interference		
EPLRS	enhanced position location reporting system	JTIDS	joint tactical information distribution system
ERP	effective radiated power	KIA	killed in action
ETD	estimated time of departure	kbps	kilobytes per second
EW	electronic warfare	kg	kilogram(s)
fax	facsimile	kHz	kilohertz
FBI	Federal Bureau of Investigation (US)	km	kilometre
		KPA	Korean People's Army
FCC	fire command centre	kW	kilowatt

LAN	local area network	PIAT	projector, infantry, anti-tank
LASER	light amplification by stimulated emission of radiation	PLA	People's Liberation Army
		PLARS	position, location and reporting system (US)
LCD	liquid crystal display	PLL	phase-locked loop
LT	lieutenant	POW	prisoner of war
LSB	lower sideband	PPI	plan position indicator
LW	longwave (150 kHz to 300 kHz)	PROM	programmable read-only memory
LZ	landing zone	PRR	personal role radio
m/s	metres per second	PSK	phase shift keying
MAJ	major	QRM	interference from another station
Mbps	megabytes per second		
medevac	medical evacuation	QRN	interference from natural sources
MFSK	multi-frequency shift keying		
MG	machine-gun	QRP	low power operation
MHz	megahertz	RA	Royal Artillery
MI	Military Intelligence	RAAF	Royal Australian Air Force
MIA	missing in action	RAE	Royal Aircraft Establishment (Farnborough)
MILSPEC	military specification		
mm	millimetre(s)	RAE	Royal Australian Engineers, Australian Combat Engineers
MMG	medium machine-gun		
MOA	minute of angle	RAF	Royal Air Force
MW	medium wave (530 kHz to 1710 kHz)	RAM	random access memory
		RAN	Royal Australian Navy
NATO	North Atlantic Treaty Organization	RAPI	reactive armour protection
		RARDE	Royal Armament Research and Development Establishment (UK)
NBC	nuclear, biological, chemical		
NCIS	Naval Criminal Investigative Service (US)		
		RAW	rifleman's assault weapon
NCO	non-commissioned officer	RCAF	Royal Canadian Air Force
NFM	narrowband frequency modulation	RCN	Royal Canadian Navy
		RCT	Royal Corps of Transport
NiCd	nickel cadmium	REME	Royal Electrical and Mechanical Engineers
NIC	network interface card		
NSW	New South Wales	RF	radio frequency
NZDF	New Zealand Defence Force	RFA	Royal Field Artillery
NZEF	New Zealand Expeditionary Force (in WWI & WWII)	RFC	Royal Flying Corps
		RFI	request for information
OIC	officer in charge	RFQ	request for quotations
OP	observation post	RFT	request for tender
PC	personal computer	RHA	Royal Horse Artillery
PCI	peripheral component interface	RLC	Royal Logistic Corps
		RN	Royal Navy
PDA	personal digital assistant	RNZAF	Royal New Zealand Air Force

Glossary of Abbreviations

RNZN	Royal New Zealand Navy	TX	transmit
ROE	rules of engagement	TX-RX	transmitter-receiver
ROM	read-only memory	UAV	unmanned aerial vehicle
RPV	remote piloted vehicle	UGV	unmanned ground vehicle
RTTY	radio teletype	UHF	ultra high frequency (300 MHz to 3 GHz)
RX	receive		
SAM	surface-to-air missile	USA	United States of America/ United States Army
SAW	squad automatic weapon		
SCSI	small computer systems interface	USAAF	United States Army Air Force
		USAF	United States Air Force
SEAL	sea/air/land	USB	upper sideband
SEATO	South East Asia Treaty Organization	USMC	United States Marine Corps
		USN	United States Navy
SEE	Signals Experimental Establishment (UK)	UTC	universal coordinated time (GMT)
SIGINT	Signals Intelligence	V	volt(s)
SINCGARS	single channel ground/air radio system	VDU	visual display unit
		VFO	variable frequency oscillator
SLR	self-loading rifle	VHF	very high frequency (30 MHz to 300 MHz)
S-meter	signal strength meter		
SMG	submachine-gun	VLF	very low frequency (15 Hz to 150 kHz)
S/N	signal to noise ratio		
snafu	situation normal all fouled up	WAC	Women's Army Corps and individual members of; obsolete
SOE	Special Operations Executive		
SOF	special operations force(s)		
SSB	single sideband	WAF	Womens' Air Force
SW	shortwave (HF band)	WAM	wideband amplitude modulation
SWR	standing wave ratio		
TCP/IP	transmission control protocol/ internet protocol	WMD	weapon(s) of mass destruction
TDMA	time division multiple access	XO	executive officer
TTG	time to go	ZULU	GMT/UTC

Appendix A: Tactical Radio Data Base

Part 1: 1900–1920

Year	Type	Wavelength in use	Country of origin
1913	Portable WS (experimental)	Unknown	GB
1914	Sterling transmitter + short-wave tuner	100–260 m	GB
	Pack set	550, 650, 750 m	GB
	Horse-drawn wagon set	800–1200 m	GB
	Wireless trench set (BF Set)	350, 450, 550 m	GB
1915	Wilson spark transmitter	350, 450, 550 m	GB
	Forward wireless: front spark set and front valve receiver	350, 450, 550 m	GB
	Forward wireless: rear spark set and rear valve receiver	350, 450, 550 m	GB
1916	Stanley CW trench set	500–1400 m	GB
	CW trench set Mk 1	500–1400 m	GB
	Tuner short-wave Mk	110–700 m	GB
1917	CW trench set Mk 1*	700–1850 m	GB
	CW trench set Mk 1**	700–1850 m	GB
	CW trench set Mk 2	340–1850 m	GB
	CW trench set Mk 3	450–1450 m	GB
	CW trench set Mk 3*	450–1450 m	GB
1918	Tuner long-wave Mk 4	3500–10,500 m	GB
	Tuner short-wave Mk 4	100–300 m	GB
1919	CW field set	2000–3000 m	GB

Appendix A: Tactical Radio Data Base

Made by	Function	Power input
Marconi	Medium wave short-distance communications	20 W estimated
Sterling	Artillery spotting from the air	30 W
Marconi	Medium wave	500 W
Marconi	Medium wave longer distance	1.5 kW
WD Factory	Front line to company	50 W
Wilson	Replacement for the BF set	130 W
WD Factory	Front line to company; forward trench communications	20 W
WD Factory	Front line to company; forward trench communications	20 W
WD Factory	Front line to company	30 W
WD Factory	Front line to company	31 W
Various	Front line to company	NA
WD Factory	Front line to company	30 W
WD Factory	Front line to company	31 W
WD Factory	Front line to company	32 W
WD Factory	Front line to company	33 W
WD Factory	Front line to company	34 W
WD Factory	Front line to company	NA
WD Factory	Front line to company	NA
WD Factory	Company to battalion	60 W

Part 2: 1921–1950

Year	Type	Frequency in use	Country of origin
1926	WT A Mk 2	2000–750 kHz	GB
1927	WT C Mk 2	500–150 kHz	GB
1933	WS No. 2	1875–5000 kHz	GB
1934	WS No. 3	1360–3333 kHz	GB
1935	WS No. 5	2.4–20 MHz	GB
1938	WS No. 7	1875–5000 kHz	GB
1939	WS No. 9	1875–5000 kHz	GB
1938	WS No. 11	4.2–7.5 MHz	GB
1939	WS No. 101	4.28–6.66 MHz	Australia
1940	WS FS6	4.28–6.66 MHz	Australia
	RX AR 7	140 kHz–25 MHz	Australia
	WS No. 8	6–9 v	GB
1941	WS 108 Mk 1	8.5–8.9 MHz	Australia
	WS 108 Mk 2	6–9 MHz	Australia
	WS 108 Mk3	2.5–3.5 MHz	Australia
	WS 109 MK2	2.5–5 MHz	Australia
	AR 12	150 kHz–15 MHz	Australia
	AR 8	140 kHz–20 MHz	Australia
	AT 5	140–500 kHz 2–20 MHz	Australia
	WS 208 Mk 1	2.5–3.5 MHz	Australia
	WS No. 19 Mk 1	2.5–6.25 MHz	GB
	Mk. VII Paraset	3–7.6 MHz	GB
	WS No. 11 (Aust.)	4.2–7.5 MHz	Australia

Appendix A: Tactical Radio Data Base 311

Made by	Function	Power output
WD Factory/ Radio Instruments	Short range telegraphy	25 W CW (input)
WD Factory/ Johnson & Phillips	Long range telegraphy	180 W CW (input)
Willis & Co.	Long range telegraphy and telephony	7 W RT 10 W CW
Ferranti	Long range telegraphy and telephony	15–400 W
Plessey RTE	Line of communications telephony	700–2000 W CW 200 W RT
SEE + unknown	Tank communications	5 W
STC	For use in AFVs	5 W RT 10 W CW
EK Cole Ltd (EKCO)	Replacement of WS No. 1 9of 1933)	0.6 W RT (l.p.) 15 W CW (l.p.) 1.5 RT (h.p.) 4.5 W CW (h.p.)
AWA	RX-TF in HF band	4W RT 6 W CW
AWA	RX-TF in HF band	4 W RT
Kingsley	Comms RX in HF	NA
Murphy Radio	RX-TF in HF band	2.5 W RT
Astor (RCA)	RX-TF in HF band	400 mW
Astor (RCA)	RX-TF in HF band	400 mW
Astor (RCA)	RX-TF in HF band	400 mW
STC	RX-TF in HF band	5 W RT 10 W CW
Astor (RCA)	Comms RX in HF	NA
AWA	Comms RX in HF	NA
AWA	Comms RX in HF	17 RW CT 50 W CW
Astor (RCA)	RX-TF in HF band	1.2–2.5 W RT 3–5 W CW
Pye Ltd	RX-TF in HF band	2.5 W RT 9 W CW
SIS and SOE	RX-TF in HF band	5 W CW
AWA	RX-TF in HF band	1.5 W RT (h.p.) 4.5 W CW (h.p.)

Part 2: 1921–1950 continued

Year	Model	Frequency	Country
1942	AT5/AR8 (also WS 112)	140–500 kHz 2–20 MHz	Australia
	RX AR 14	111 kHz–15 MHz	Australia
	3BZ Teleradio (C6770)	2.5–10 MHz	Australia
	3BZ Teleradio	2.5–10 MHz	Australia
	ATR 2B	3–7.5 MHz	Australia
	ATR 2C	3–7.5 MHz	Australia
	ATR 4A	3–7.5 MHz	Australia
	ATR 4B	3–7.5 MHz	Australia
	RX Set No. 1	480 kHz–26 MHz	Australia
	WS RC 8	3–10 MHz	Australia
	RX C 7000	60 kHz–30 MHz	Australia
	BC1000	40–48 MHz	USA
	Type 3 Mk. 1 'B1'	3.8–15.8 MHz	GB
	Type 3 Mk. 2 'B2'	3.1–15.2 MHz	GB
	RX Set No. 4	1.2–20 MHz	Australia
1943	WS 109 Mk 1	2.5–5 MHz	Australia
	WS 19 Mk 2 (Aust.)	2–8 MHz	Australia
	AMR 100	100 kHz–27 MHz	Australia
	WS 108 Mk 3	2.5–3.5 MHz	Australia
	AR 17	100–150 MHz	Australia
1944	WS DR 106	60–80 MHz	Australia
	AMR 101	100 kHz –26 MHz	Australia
	AMR 300 (also A679H)	1.5–24 MHz	Australia
	WS 208 Mk 2	2.5–3.5 MHz	Australia
	WS No 22 (Aust.) Yellow	2–8 MHz	Australia
	Type A Mk 3	3.2–5.2 MHz	GB
	WS No 122 (Aust.)	2–8 MHz	Australia
1945	AMR 300 (A 679K)	1.5–24 MHz	Australia
	AR 88	535 kHz –32 MHz	USA
	WS 88	38 –42 MHz	GB
	C/PRC 26	47–55.4 MHz	Canada
	WS 62	1.6–10 MHz	Australia
	PRC 8	20–27.9 MHz	USA
	PRC 9	27–38.9 MHz	USA
	PRC 10	38–55 MHz	USA
	BC-611	3.5–6 MHz	USA
1947	WS 31	40–48 MHz	GB
	WS 128 Mk 1	2–4.5 MHz	Australia

Appendix A: Tactical Radio Data Base

AWA	RX-TF in HF band	17 W RT
		50 W CW
Astor (RCA)	Comms RX in HF	NA
AWA	Comms RX in HF	NA
AWA	Comms RX in HF	10 W RT
		12 W CW
Astor (RCA)	RX-TF in HF band	15 W CW
Astor (RCA)	RX-TF in HF band	15 W CW
Astor (RCA)	RX-TF in HF band	2 W CW
Astor (RCA)	RX-TF in HF band	2 W CW
Kingsley	Comms RX in HF	NA
Astor (RCA)	RX-TF in HF band	10 W CW
AWA	Comms RX in HF	NA
Galvin	VHF FM	0.3 W
SOE	RX-TF in HF band	9–18 W CW
SOE	RX-TF in HF band	20 W CW
Phillips	Comms RX in HF	NA
STC	RX-TF in HF band	5 W RT
		10 W CW
AWA	RX-TF in HF band	6 W CW
AWA	Comms RX in HF	NA
Astor (RCA)	RX-TF in HF band	600 mW
Astor (RCA)	Comms RX in HF	NA
Phillips	CHF	20 W RT
AWA	Comms RX in HF	NA
STC	Comms RX in HF	NA
Astor (RCA)	RX-TF in HF band	400 mW CW
Astor (RCA)	RX-TF in HF band	2.5 W RT (h.p.)
		10 W CW (h.p.)
		1 W RT (l.p.)
		1.5 W CW (l.p.)
SOE	RX-TF in HF band	5 W CW
Astor (RCA)	RX-TF in HF band	1 W RT
		1.5 W CW
STC	Comms RX in HF	NA
RCA	Comms RX in HF	NA
Many	VHF FM	250 mW
Rogers	VHF FM	300 mW
Phillips	HF CW AM	1.1 W CW
		0.8 W RT
Many	VHF FM phone	1.2 W
Many	VHF FM phone	1.0 W
Many	VHF FM phone	0.9 W
Galvin	HF AM Handie-Talkie	36 mW
Murphy Radio	VHF FM backpack	300 mW
Tasma	RX-TX HF backpack	250 mW

Part 3: 1951–1975

Year	Type	Frequency in use	Country of origin
1952	WS 128 Mk 2	2–4.5 MHz	Australia
1955	WS A510	2 –10 MHz	Australia
	C42	36–60 MHz	GB
	C34	23–38 MHz	GB
1960	AN/PRC 47	2–12 MHz	USA
	AN/PRC 25	30–75.95 MHz	USA
1962	PRC F1	2–11.9 MHz	Australia
	AN/PRC 77	30–75.95 MHz	USA
	PRT 4 PRR 9	47–57 MHz	USA
1965	GRC 109	3–24 MHz	USA
	RS 6	3–24 MHz	USA
	AN/VRC 12	30–76 MHz	USA
	AM/TRC 75	2–20.99 MHz	USA
	C11/R210	2–16 MHz	GB
	AN/GRC 106	2–30 MHz	USA
	AN/GRC 125	30–75.95 MHz	USA
	AN/PRC 64A	2.1–5 MHz	USA
	R391/URR	500 kHz–32 MHz	USA

Appendix A: Tactical Radio Data Base

Made by	Function	Power output
Tasma	RX-TX HF backpack	250 mW
AWA	RX-TX RT CW manpack	760 mW CW 138–245 mW RT
Plessey	VHF FM	15 W(h.p.) 0.25 W (l.p.)
Unknown	VHF FM	15 W (h.p.) 0.25 W (l.p.)
Collins	HF SSB CW ground	20–100 W
RCA + many	VHF FM manpack	1–1.5 W
AWA	HF SSB manpack	5 W CW 10 W SSB
RCA + many	VHF FM manpack	1–1.5 W
Delco	VHF FM squad	0.15–0.5 W
Unknown	Clandestine	12 - 15 W
Unknown	Clandestine	12 - 15 W
Associated Industries	VHF FM vehicle-mounted	35 W (h.p.) 8 W (l.p.) 100 W
Collins	HF SSB CW Duplex TTY	50 W (h.p.)
British Military	HF AM CW FSK	5–10 W (l.p.) 200 W CW FSK
General Dynamics	AM SSB RATT CW	400 W USB 50 W tune 1.1–2 W FM
RCA + many	VHF FM vehicle mounted	5 W CW
Delco	AM CW high speed patrol	NA
Collins	HF SSB CW receiver	

Part 4: 1976-2012

Year	Type	Frequency in use	Country of origin
1976	AN/PRC 47	2-12 MHz	USA
	AN/PRC 77	30-75.95 MHz	USA
	AN/PRC 68	30-79.95 MHz	USA
	PRC F1	2-12 MHz	Australia
1986	AN/PRC 119 RT 1439	30-88 MHz	USA
1990	RT F 100 Raven	2-30 MHz	Australia
	RT F 200 Raven	30-88 MHz	Australia
	RT F 700 Pintail (AN/PRC 6725E)	30-470 MHz	Australia
1995	AN/PRC 6725E	30-88 MHz	USA
	AN/PRC 139	30-88 MHz 136-174 MHz 403-470 MHz	USA
2000	AN/PRC 148 MBITR	30-512 MHz	USA
2005	AN/PRC 343	2.4-2.483 GHz	GB
2012	AN/PRC 152	30-152 MHz	USA
	AN/PRC 150	1.5-59.995 MHz	USA
	EPLRS-XF RT 1720	225-450 MHz	USA
	Microlight RT 1922	225-2000 MHz	USA

Appendix: Tactical Radio Database

Made by	Function	Power output
Collins	Higher power HF SSB	20–100 W
RCA	VHF FM encrypted	1–1.5 W FM
Magnavox	Tactical hand-held	1–1.5 W FM
AWA	Back pack HF SSB CW	10 W PEP SSB 5 W CW 1 W all modes
Magnavox	Tactical hand-held	1–4 W FM
Plessey Siemens	HF multi-mode	20 W SSB
Plessey Siemens	VHF multi-mode	5 W FM
Thales	VHF FM hand-held	0.5–2 W FM
Racal	Hand-held	0.5–2 W
Thales	Hand-held	0.5–2 W
Thales	Hand-held squad multi-mode	0.1–5 W AM/FM
SELEX	PRR spread-spectrum UHF	100 mW
Harris	VHF UHF multi-mode hand-held	250 mW–5 W
Harris	HF multi-mode backpack	1, 5, 20 W PEP SSB/FM
Raytheon	LAN mobile node UHF	400 mW, 3 or 20 W select
Raytheon	Squad multi-function	100 mW–4 W

Appendix B:
Larger Schematic Diagrams

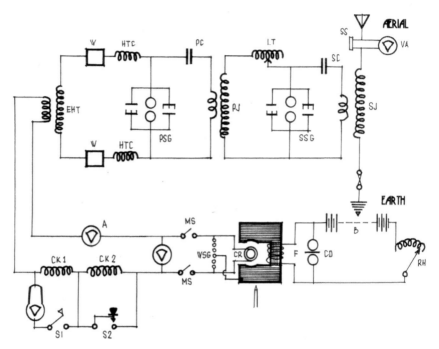

A	ALTERNATOR FRAME	SSG	SECONDARY SPARK GAP WITH PROTECTORS
CR	ALTERNATOR COLLECTOR RINGS	SJ	SECONDARY JIGGER
B	BATTERY	CK 1, CK2	SIGNALLING CHOKES
RH	RHEOSTAT	S1,S2	SIGNALLING SWITCHES
CD	CARBON DISK PROTECTOR ACROSS FIELD WINDINGS	EHT	TRANSFORMER 2000 - 20,000 VOLT RATIO
F	ALTERNATOR FIELD WINDING	HTC	HIGH TENSION INDUCTANCE
WSG	WURTZ SPARK GAP PROTECTOR ACROSS ARMATURE	WW	WURTZ ARRESTORS
MS	MAIN SWITCH	PC	PRIMARY CONDENSER
V	VOLTMETER	PSG	PRIMARY SPARK GAP
A	AMMETER	PJ	PRIMARY JIGGER
LT	ADJUSTABLE TUNING INDUCTANCE	VA	AERIAL AMMETER
SC	SECONDARY CONDENSER	SS	AMMETER SHUNT RESISTANCE

Original transmitter at Poldhu

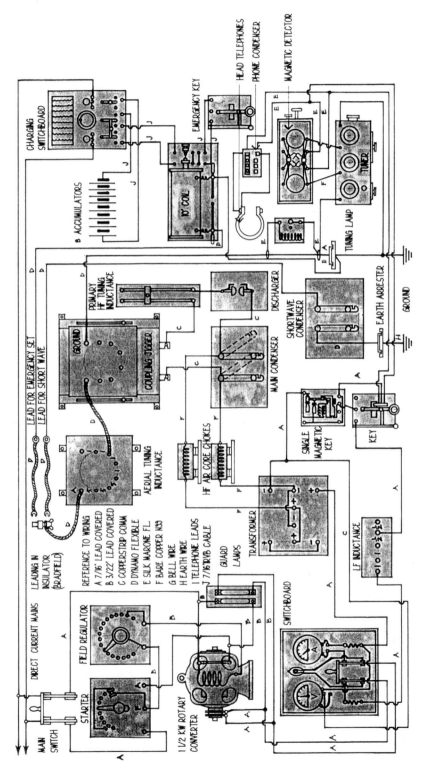

Schematic diagram of typical marine wirelss station, 1912

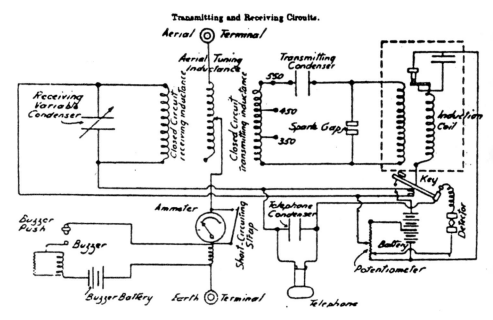

Schematic for the BF trench set

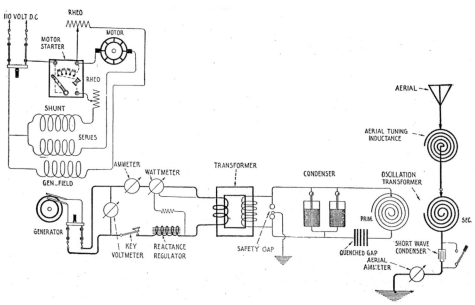

Marine quenched spark transmitter, 1916

Appendix B: Larger Schematic Diagrams

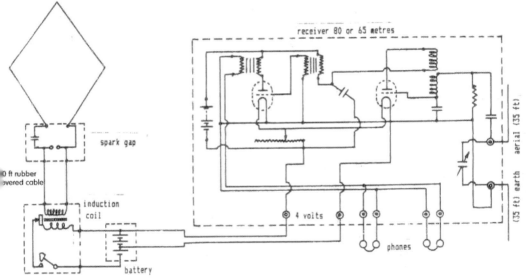

Schematic of forward loop set

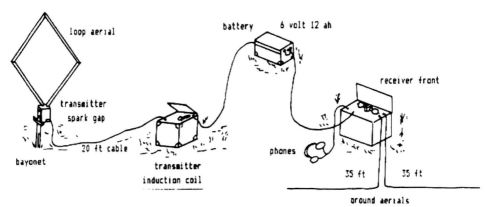

Forward spark transmitter–valve receiver with aerials

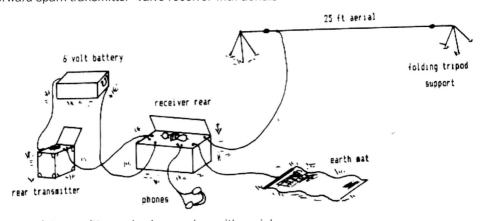

Rear spark transmitter and valve receiver with aerial

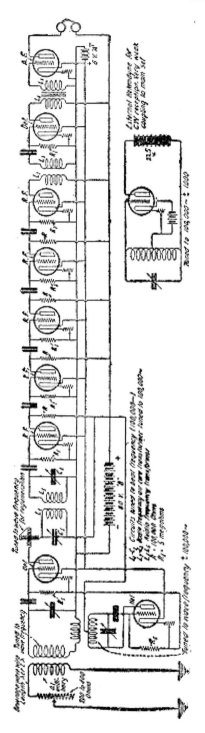

Armstrong superheterodyne receiver—first amateur trans-Atlantic contact (WW).

Appendix B: Larger Schematic Diagrams 323

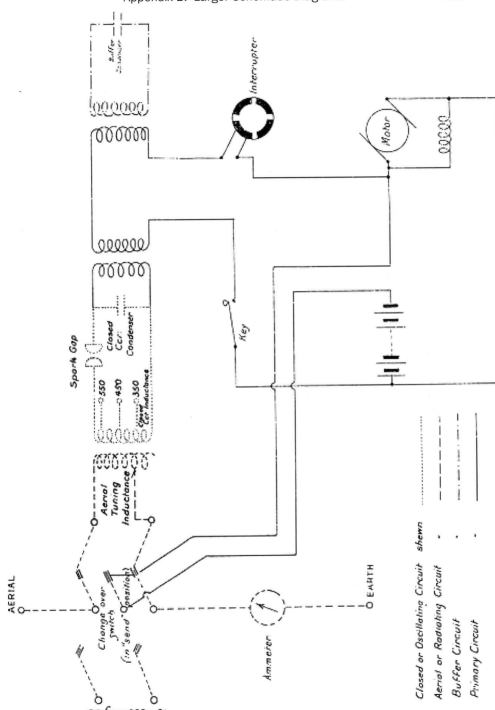

Schematic for the Wilson transmitter (LM).

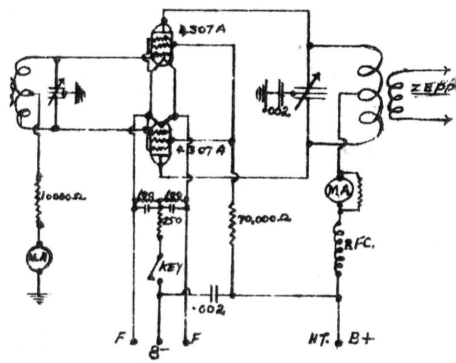

Above: Winnie the War Winner RF amplifier schematic (PRJ).
Below: Cork set receiver (KMM)

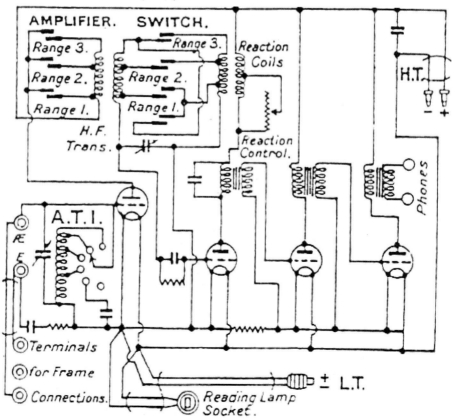

Appendix B: Larger Schematic Diagrams

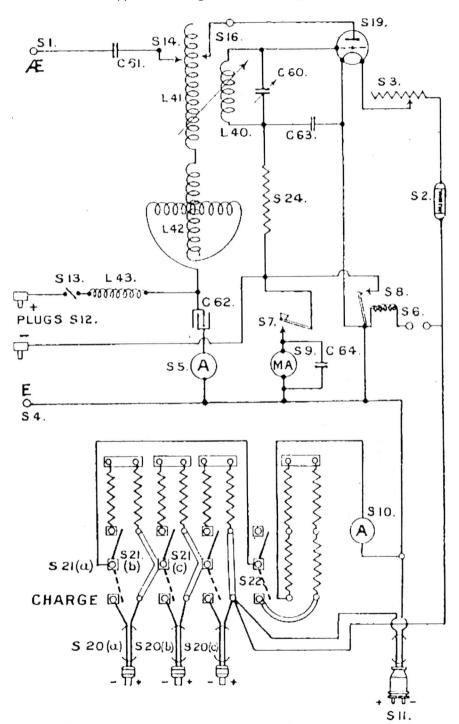

Cork set transmitter (KMM).

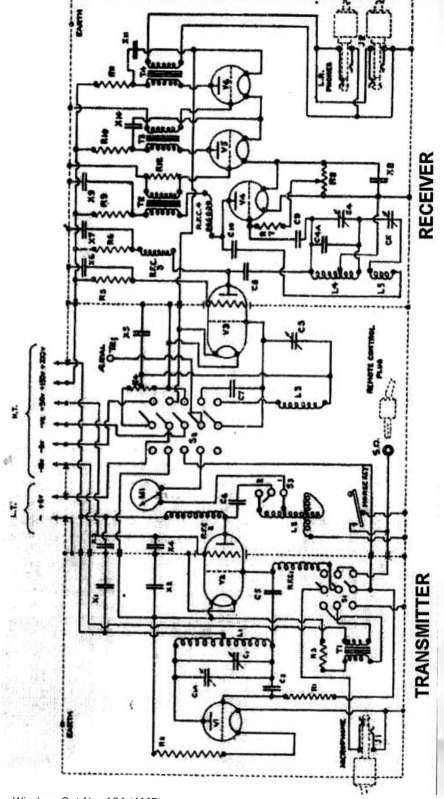

Wireless Set No. 101 (AMP).

Appendix B: Larger Schematic Diagrams

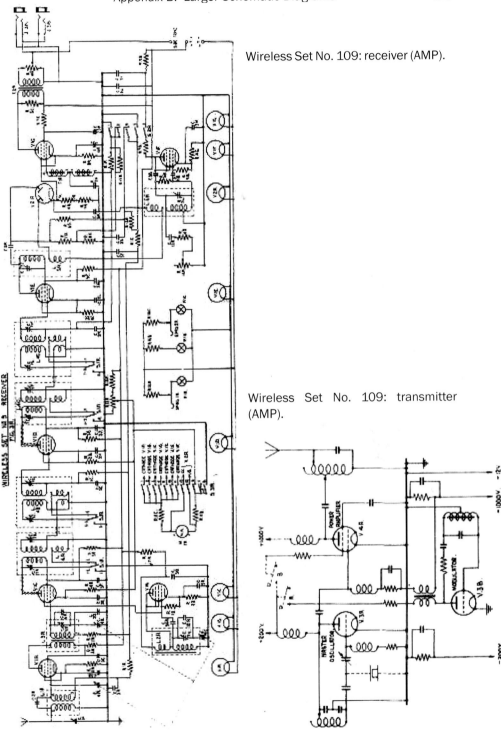

Wireless Set No. 109: receiver (AMP).

Wireless Set No. 109: transmitter (AMP).

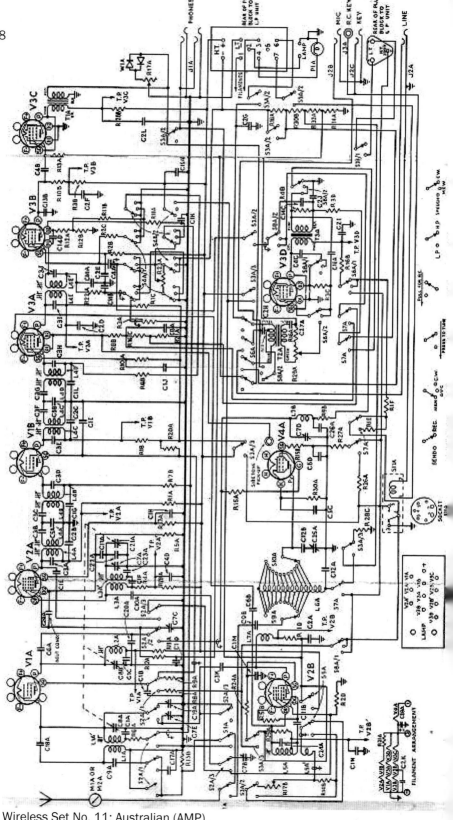

Wireless Set No. 11: Australian (AMP).

Appendix B: Larger Schematic Diagrams

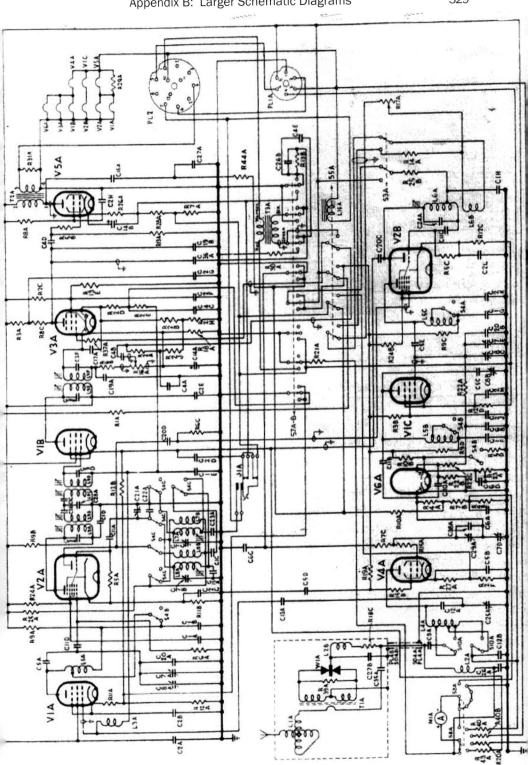

Wireless Set No. 19 (Aust.) (AMP).

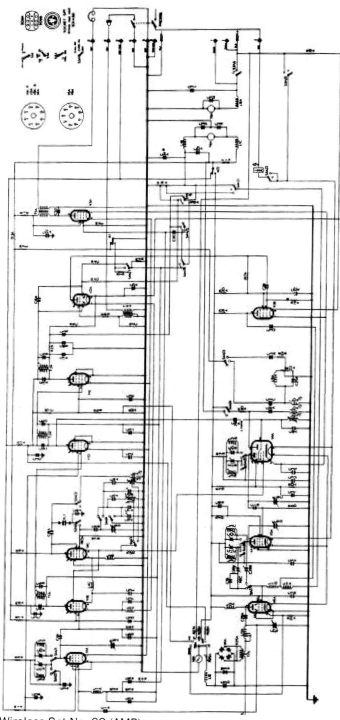

Wireless Set No. 62 (AMP).

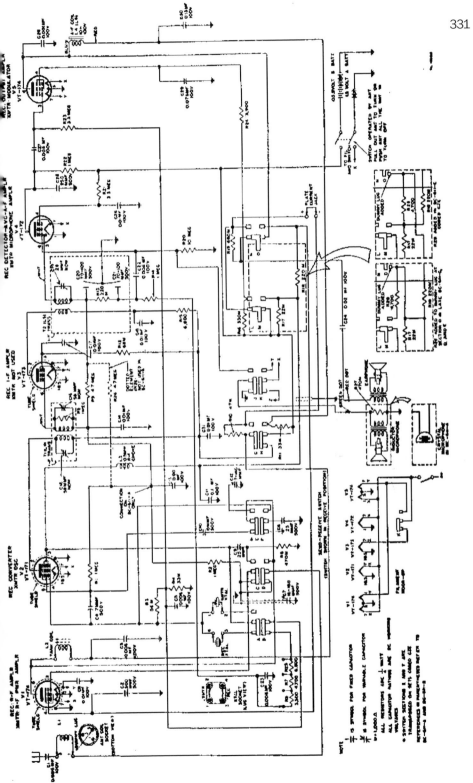

BC-611: the Handie-Talkie (AMP).

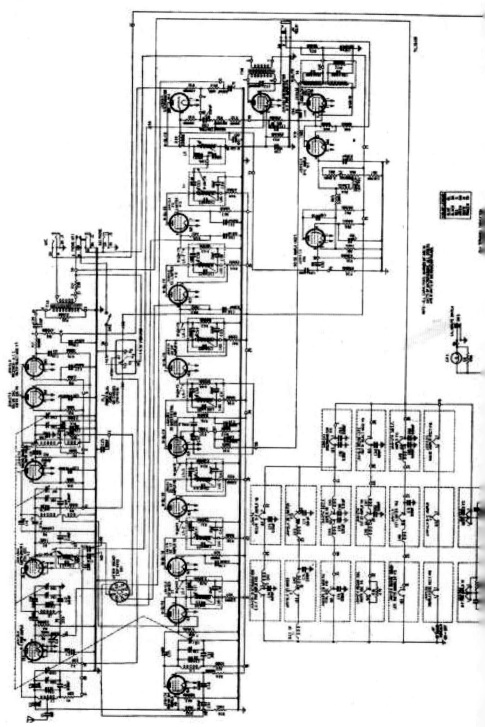

BC-1000 transceiver: the Walkie Talkie (AMP).

Appendix B: Larger Schematic Diagrams

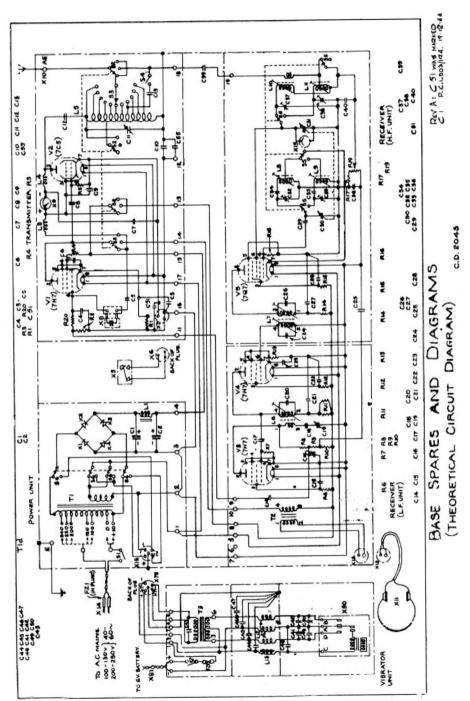

Type A Mark 3 clandestine transceiver (AMP).

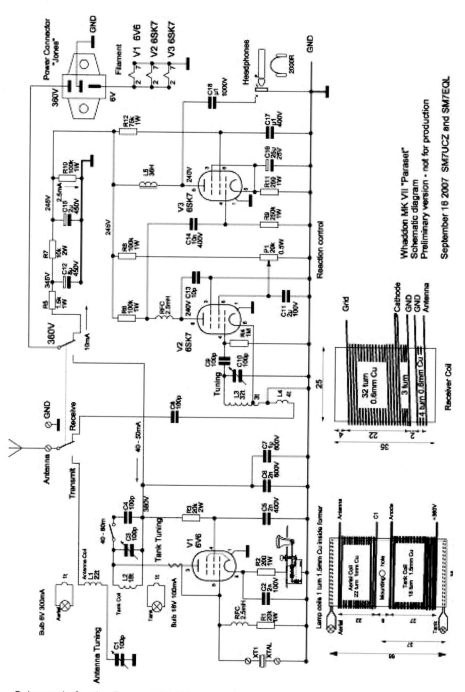

Schematic for the Paraset (SM7).

Appendix B: Larger Schematic Diagrams

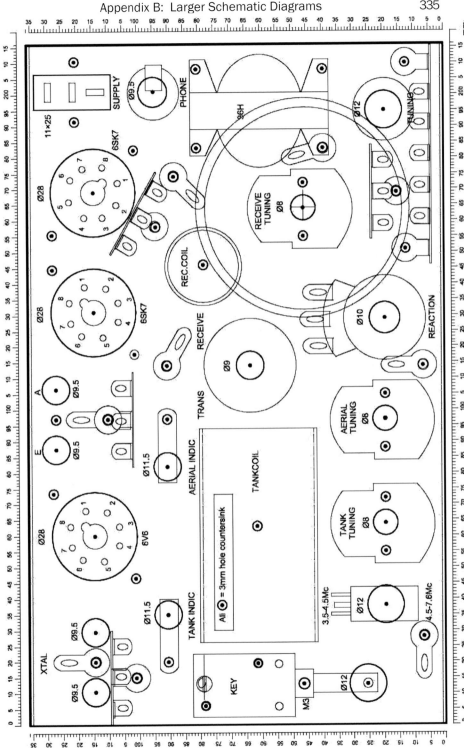

Top panel layout for the Paraset (SM7).

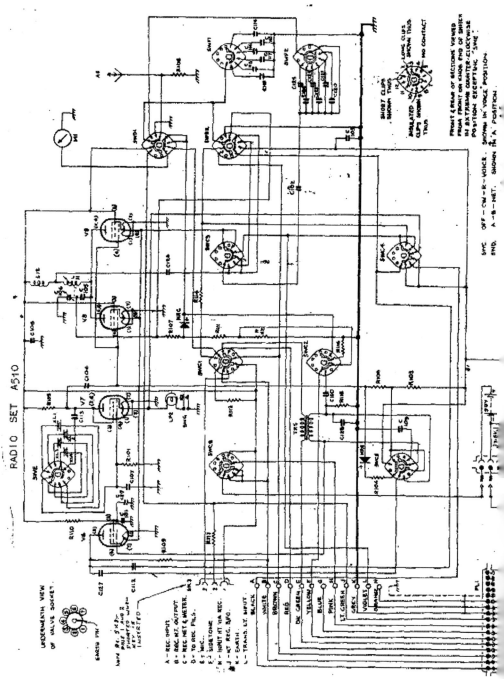

Wireless Set A510: transmitter section (AMP).

Appendix B: Larger Schematic Diagrams

Wireless Set A510: receiver section (AMP).

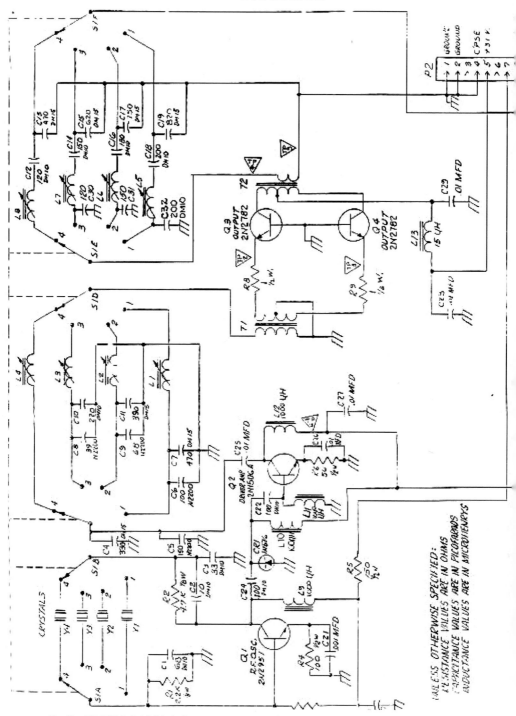

Radio AN/PRC-64A high frequency transmitter board (AMP).

Appendix B: Larger Schematic Diagrams

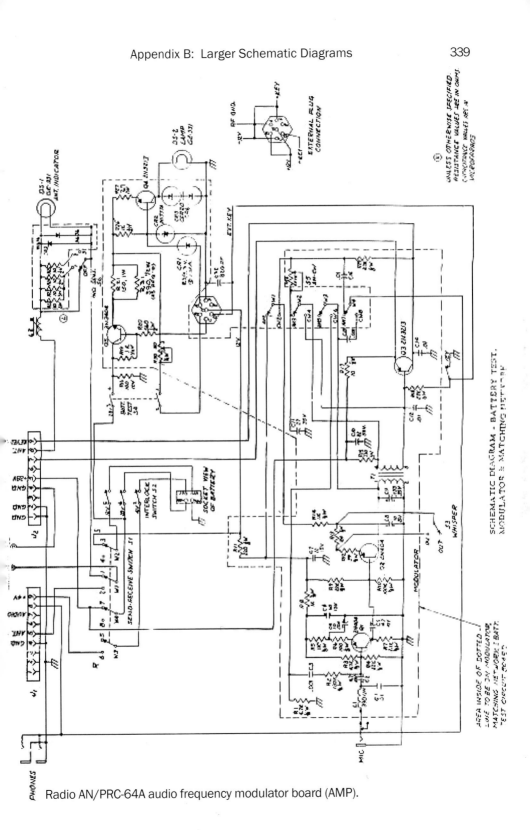

Radio AN/PRC-64A audio frequency modulator board (AMP).

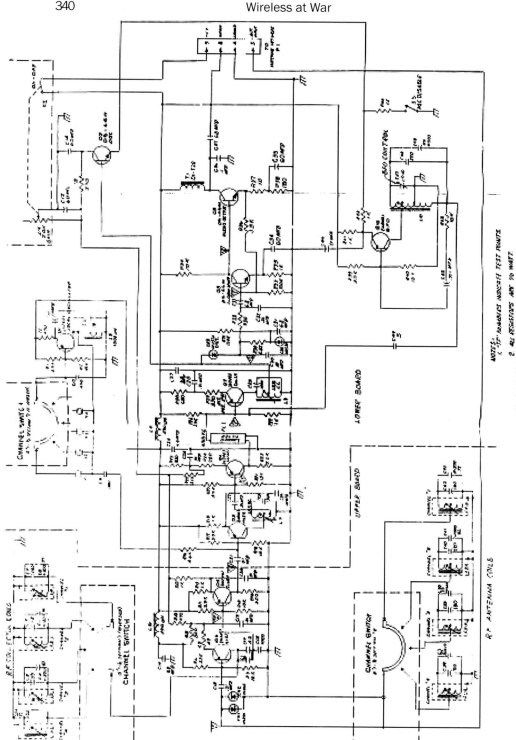

Radio AN/PRC-64A receiver board (AMP).

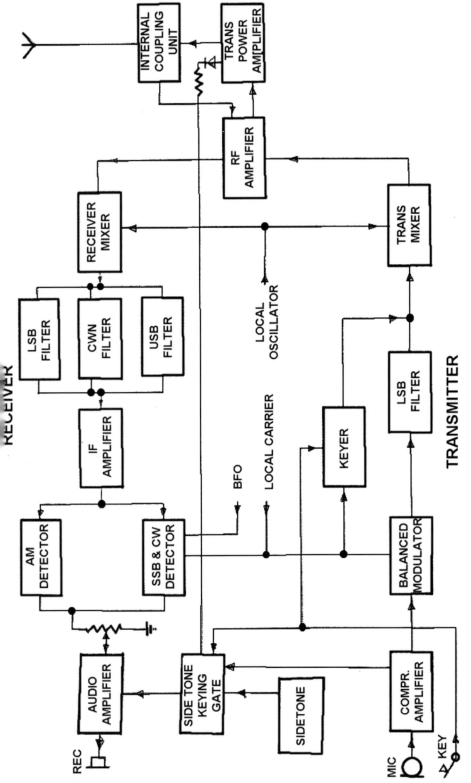

PRC-F1 block diagram (AMP).

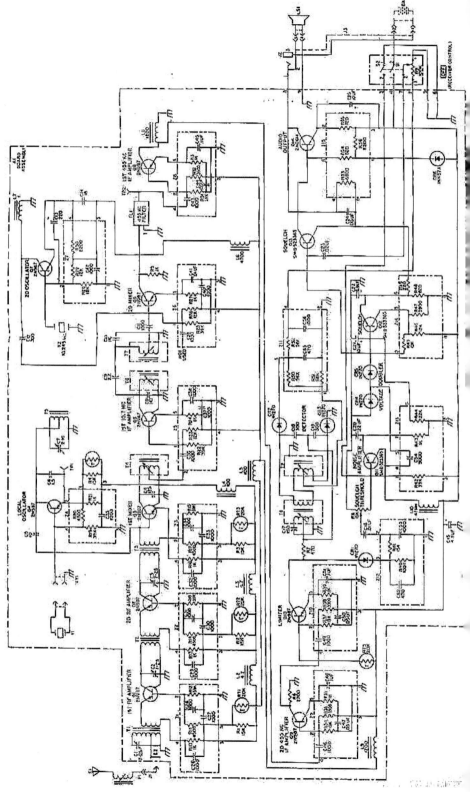

AN/PRR-9 VHF FM receiver, helmet mounted (PRJ).

Appendix B: Larger Schematic Diagrams

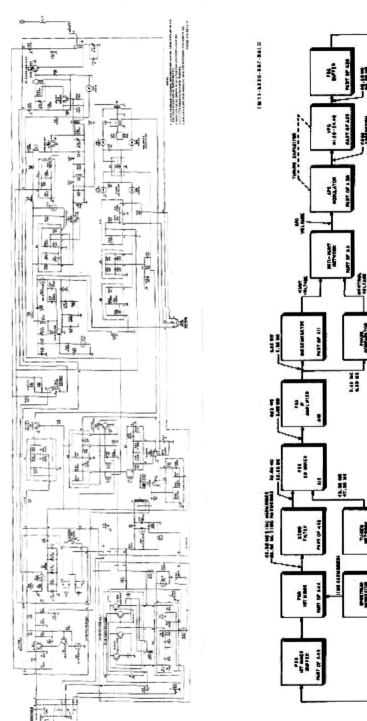

AN/PRT-4 VHF FM transmitter, hand-held (PRJ).

AN/PRC-77 VHF FM backpack radio synthesiser section.

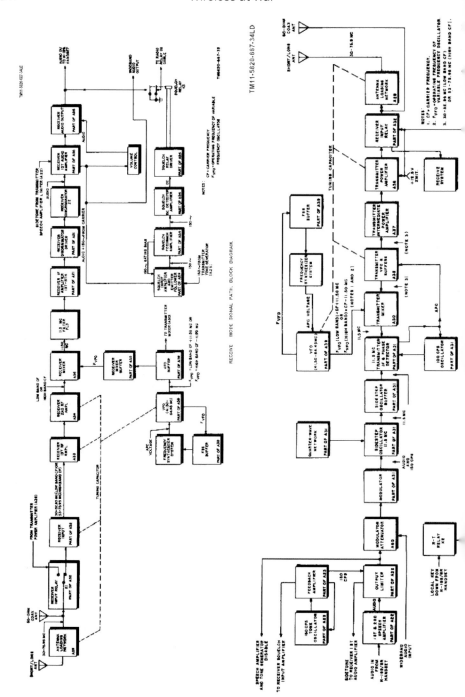

AN/PRC-77 VHF FM backpack radio schematic (AMP).

Appendix B: Larger Schematic Diagrams

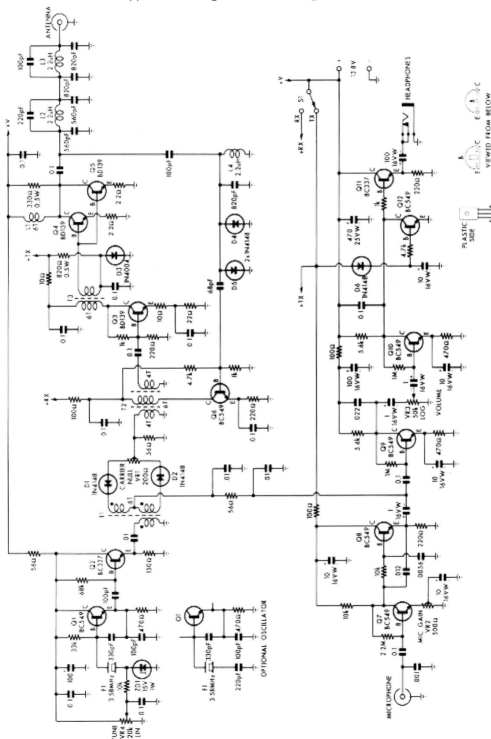

Circuit diagram of low-power double-sideband transceiver (SCM).

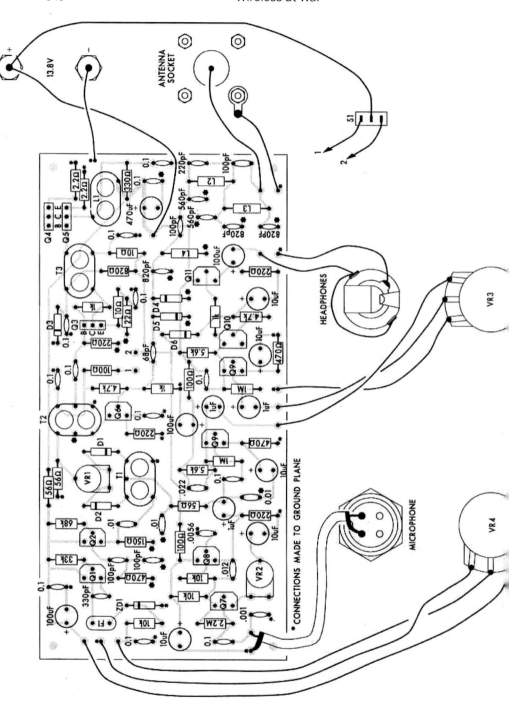

Schematic layout of circuit board and external wiring (SCM).

Index

A
A510 · 201, 219, 229–30, 231, 314
Adelaide, HMAS · 252
Admiralty, British · 34, 56, 125
Adrianople · 38
aerial · 44
Afghanistan · 11, 251, 254, 276, 281
Air Defense Systems Engineering Committee (US) · 206
airship · 112
Airzone Cub · 109
al-Qaeda · 11, 254
Alexanderson, Ernst · 83–4
Allenby, Field Marshall Edmund · 65
Allies · 81, 107, 128–9, 144
Alsace-Lorraine · 107
Altair 8800 computer · 226
alternator · 83
alternator transmitter, Grimeton· 84
Amalgamated Wireless (Australasia), AWA · 75, 115, 133, 136, 154, 155, 201, 220, 229, 234, 239, 242, 311, 313, 315, 316
Amalgamated Wireless Valve · 115, 136, 169
amateur radio · 102, 143
Amberley Chalkpits Museum · 7
American isolationism · 108
Amstrad microcomputer · 248
AN/MTC-1 · 263
AN/PRC-25 VHF · 216
AN/PRC-47 · 218, 273, 274
AN/PRC-64 · 219, 220, 231–2
AN/PRC-64A · 219, 231–4
AN/PRC-77 · 219, 220, 237–9, 266, 267, 269, 272, 273, 275, 282, 290
AN/PRC-148 MBITR · 277, 278
AN/PRC-150 · 282–3
AN/PRC-152 · 281–2
AN/PRC-343 · 279
AN/TRC-75 · 217–8

AN/TSQ-158 · 283
AN/UGC 144 · 264
Android · 265
Ankara · 129
Antheit, George · 269
appeasement · 108, 125
Apple computer · 226
Arab army · 65
Archduke Ferdianand, assassination of · 45
Arco, Count Georg von · 14
Ardennes · 127
Armstrong, Edwin Howard · 93-4, 109, 161
Armstrong receiver · 94
Arnhem, Battle of · 136, 170, 173, 290
ARPA · 208
ARPANET · 208–9, 211, 226
Arras · 73
ASCII · 264, 270
Astor · 311, 313
asymmetric warfare · 210, 251–2
atomic bomb · 132, 143, 189
Audion (valve) · 73, 88–9, 93, 109
AuSteyr F88 rifle · 281
Austin, Brian · 32, 151
Australia · 11, 59, 75, 126, 127–8, 129–32, 140, 143, 153–9, 192–4, 206, 215–16, 251–2, 253, 254, 257, 272
Australia and New Zealand Army Corps, ANZAC · 60, 63, 68
Australia New Zealand United States Treaty, ANZUS · 194, 206, 251
Australian Broadcasting Commission, ABC · 58
Australian Defence Force, ADF · 280
Australian Defence Industries, ADI · 281
Australian Imperial Force, AIF · 57, 60

Australian War Memorial · 8, 46, 58, 157, 159
automatic link establishment, ALE · 283
AWA panel set · 76
AWA Teleradio 3B · 154
Ayesha (yacht) · 54
Azerbaijan · 66
B
Baghdad · 66, 69
Baird, John Logie · 110
Baird Televisor, 1929 · 110
Baku · 66, 68
Balkans, the · 38
Baltic Sea · 36, 51
Bao Dai · 213
Barack Obama, President · 254
Baran, Paul · 207
Basra · 66
Batavia · 54
Battle of Britain · 76, 111, 127
Baudot code · 226, 264
Bawdsey Manor · 114
Bazna, Eleysa · 129
BC-1000 Walkie Talkie · 137, 138
BC-611 Handie-Talkie · 137, 164, 173
Beard, Maston · 224
beat frequency oscillator, BFO · 166
Belgium, invasion of · 55, 127
Berlin, capture of · 132
BF trench set · 2, 70–2, 73, 85, 95, 96, 163
Billings, Bert · 57–8, 61
bin Laden, Osama · 254
Binns, Jack · 39
Black Box · 16, 20–1
Blamey, General Thomas · 131
Blandford Forum Museum · 96, 97
Bletchley Park · 122–3, 204, 224, 228
Blitzkrieg · 119
Boer War· 20, 29–34, 42
Bolt Beranek & Newman · 207, 208

347

Boot, Harry · 113
Borneo campaign · 131
Bovington Tank Museum · 8
Branly, Edouard · 17, 87, 222
Braun, Professor Karl · 14, 89
Brean Down, UK · 18
Brest-Litovsk, Treaty of · 108
Bristol Channel · 18
British Broadcasting Commission, BBC · 124
British Vintage Wireless Society, BVWS · 177
Brunei campaign · 131
Buddhists, persecution of · 213
Bulgaria · 38, 108
Buller, Major-General Sir Redvers · 32, 33
Buna · 131, 132
Burke, Eric Keast · 68
Burma campaign · 118
Bush, President George W · 253, 254
Bushmaster vehicle · 285

C
C/PRC-26 VHF · 197
C11-R210 HF · 203
Cabot Tower, St Johns, Newfoundland · 26
Caernarvon wireless station · 85
Cairo · 59, 129
Californian, SS · 39
Calzecchi-Onesti, Temistocle · 87
Cambrai, Battle of · 79
Canada · 314
Canungra Jungle Warfare training camp · 198
Carpathia, SS · 39
Carr, Major · 14
Caspian Sea · 65, 67
central processing unit, CPU · 264, 265
Cerf, Vinton · 209
Chamberlain, Neville · 124, 125, 126
Chauvel, General Sir Harry · 60
Chelmsford · 9, 37
Chiang Kai-shek · 190
Childers, Erskine · 77
China · 191
Chin Peng · 198
Churchill, Winston Spencer · 56, 79, 125, 132, 141, 189
Citizen Military Forces · 131, 201

Clausewitz, Carl von · 210
Clifden Wireless Station · 85
coastwatchers · 153–4
Cocos-Keeling Islands · 53
codebreaking · 121–3, 204–5, 224
Cold War · 205–9
Collins · 315, 317
Columbus's egg · 17
commercially developed communications apparatus, COTS · 271, 280
Commonwealth Scientic and Industrial Research Organisation, CSIRO · 224
Communism · 138, 189
Communist China · 191, 192, 193
continuous wave, CW · 73–5, 136, 142, 165–6, 167, 174, 182, 201, 202, 219, 230, 232–3, 234, 248, 274, 282, 311, 315, 317
Coral Sea, Battle of the · 130
Cornwall · 16, 18
Cosgrove, General Peter · 253
Cowes, UK · 17
Crete campaign · 129
CSIRAC computer · 224
Curtin, John · 130
Czechoslovakia, annexation of · 125

D
Damascus · 65
Dardanelles campaign · 64, 65, 66
Darwin, attack on · 130
Davies, Donald · 208
D-Day, 6 June 1944 · 266
De Forest, Lee · 88–9, 93, 109
De Mole, Launcelot Eldin · 45–9, 78
Delagoa Bay, blockade of · 34
Dien Bien Phu · 206, 213
dispersed network · 211
Dobbyns, Bill · 58, 61
Dogger Bank battle · 36
Domville, Vice-Admiral Barry · 19
Donovan, John · 158
Dowding, Air Chief Marshall Sir Hugh · 76
Dreadnought, class of battleship · 51, 79
Dunkirk · 127
Dunsterforce · 66–9

Dunsterville, General Lionel 'Stalky' · 66–9
Dunwoody, General Henry · 90
Durban, blockade of · 34
Dutch Provincial Reconstruction Force · 254
Duxford Museum · 8

E
East Timor · 253
Eckert, J Presper · 224
Edison, Thomas · 88, 160
Edison valve · 88, 160
Egypt · 59, 65, 69
Einstein, Albert · 144–5
Eisenhower, General Dwight D · 207
EKCO radio UK · 110, 311
El Alamein, Second Battle of · 130
Elbit Battle Management System · 260, 281, 290
Elbit Industries · 281, 290
electronic counter-counter measures, ECCM · 283
electronic counter measures, ECM · 284
Elgar-Whinney, John · 7
Emden, SMS · 52–4
electromotive force, EMF · 88
EMI Ltd · 111
encryption · 265–8
Enewetak Atoll · 189
enhanced positiion and location reporting system, EPLRS · 283, 285
ENIAC computer · 224
Enigma codes · 119–22, 129, 265
Estonia · 108
Euphrates River · 66

F
Falkenberg, Bengt · 176, 179
Falkenhorst, General Niklaus von · 148
Falkland Islands · 52
Falklands War · 255–7
Fanning Island · 52
Faraday, Michael · 12–13
Fareham Naval Radio Museum · 75
Farnsworth, Philo · 111
Ferranti · 278, 311
Fessenden, Reginald · 82, 83
Finland · 108
Finmeccanica · 278
Flanders, trenches of · 50, 288

Index

Flatholm Island · 18
Fleming valve · 89
Fleming, Professor J Ambrose · 26, 86, 88, 222
Florida, SS · 38, 39
forward spark transmitter · 74, 91
forward valve receiver · 74
France, surrender of in WW2 · 51
Franklin, Charles S · 83
frequency hopping · 266–8, 277
frequency shift keying, FSK · 283
Friesian Islands · 77
Fuller phone · 63

G

Gallipoli campaign · 30, 56–63, 69, 290
Gallipoli wireless station · 61
Geheimshreiber · 204
Gel-cell battery · 183, 185, 245
General Electric Corporation, GEC · 277
General Electric, GE · 83, 89, 207
German Army · 70, 73
Gloster Meteor fighter · 193
Gneisenau, SMS · 142
Godley, Paul · 94
Gordon, Major Jim · 7
Government Code and Cypher School · 122, 204
GPS · 289, 290
Great Britain · 7, 10, 11
Grimeton radio station · 83–4
Guderian, General Heinz · 119, 120

H

Habibie, President · 253
Hahn, Otto · 143
Hamel, Battle of · 79
Handie-Talkie · 137, 163–4, 172, 194, 290, 312
Hardanger plateau, Norway · 146, 147
Harris Corporation · 276, 281, 282
Haugland, Knut · 147, 148–51
Haukelid, Knut · 147, 152
heavy water · 143, 145–6
Helberg, Claus · 147
Heroes of Telemark, The · 147
Hertz, Heinrich · 10, 12, 13, 14, 21

Hertz oscillations · 15
Hindenburg Line · 73
Hirohito, Emperor · 132
Hiroshima · 132, 189
Historic Radio Society of Australia, HRSA · 105, 177
Hitler, Adolf · 51, 107, 119, 125, 126, 127, 132, 144
Honeywell · 208
Ho Chi Minh · 213
Ho Chi Minh Trail · 213
horse-drawn Marconi wireless · 56

I

IBM · 207
Iconoscope camera · 111
Incheon, Battle of · 191
Indochina · 206, 212–3
Indonesia · 131, 210, 253
Information Processing Techniques Office, IPTO · 207
Inter-Allied Services Department, IASD · 155
Internet · 206, 208–9, 261, 271
internet protocol, IP · 209
Iran · 250
Iraq · 251
Iraq War, Phase 1 · 252
Iraq War, Phase 2 · 253–4
Iron Curtain · 132, 138
ironclads · 78
Isandlwana · 288

J

Jackson, Captain Henry · 14, 18, 19, 75
Jameson-Davies, Henry · 15
jamming · 267
Japan · 34–5, 36, 37
jeep-mounted wireless · 134
Jerusalem · 65
Johannesburg · 29
Joliot, Jean Frédéric · 143, 144
Journeaux, Bill · 7
jungle communications · 199–210
Juno, HMS · 18

K

Kahn, David · 121
Kapyong, Battle of · 193
Kazakhstan · 189, 205
Kennedy, President John F · 213
Kettering ignition system · 100
Kilby, Jack · 225
Kim Il-sung · 192

Kim Jong-il · 192
Kim Jong-un · 192
Kimberley mobile wireless stations · 42
Kimberley mobile wireless stations · 33
Kimura, Professor · 37
Kingsley radio · 311, 313
Kjelstrup, Arne · 147
Knatchbull-Hugesson, Sir Hughe · 129
Koepang · 156
Kokoda Track · 131–2
Korean War · 190–7, 210, 212
Krait (see Operation Jaywick)
Kruger, President Paul · 31, 32
Kurrajong Military Radio Museum · 8
Kut-al-Amara, Iraq · 66

L

Ladysmith, siege of · 32, 42
Lamarr, Hedy · 267
Land Warfare Studies Centre · 258
Langmuir, Irving · 89
Larkspur radios · 194, 201–3
Latvia · 108
Lawrence, TE (of Arabia) · 65
Lee-Enfield rifle · 281
Levy, Lucien · 94
Liaodong Peninsula · 35
Licklider, JCR · 207
Lithuania · 108
local area mobile network, LAN · 289
Lodge, Professor Sir Oliver · 12, 13, 17, 22, 82, 110, 222
Long Tan, Battle of · 215
Lorenz encoding machine · 204
Los Alamos Atomic Research Centre · 132, 189
Loveless, Max · 158
Ludendorff, General Eric · 79
Luftwaffe · 111, 114, 122
Lyon, Lieutenant-Colonel Ivan · 155

M

MacArthur, General Douglas · 131–2, 191
Mafeking, seige of · 42
Magicienne, HMS · 34
Magnavox · 238, 317
magnetic detector ('Maggie') · 20, 86–8
Malay Archipelago · 131
Malayan Emergency · 197–9

Manchuria, Russian invasion of · 35-6
Manhattan Project · 143
Maori Wars · 30
Marconi, Guglielmo · 12, 13, 14-18, 20, 26, 75, 82, 85, 87, 89, 222
Marconi (Company) · 15, 24, 26, 32, 33, 37, 41, 45, 75, 83, 88
Marconi radio equipment · 19, 22, 24, 32, 33, 34-5, 39-40, 55, 56, 57-9, 61, 67, 68, 69, 70, 75, 81, 83, 84, 85, 86, 87, 95, 98, 105, 309
Marconi coherer of 1902 · 24
Marconi coherer receiver of 1902 · 25
Marconi crystal set · 84
Marconi pack set · 57, 59, 61, 68, 67, 84, 85, 95
Marconi wagon-pack set · 38
Marne, Battle of the · 55, 56, 81, 95
Marshall Islands · 55, 189
Maryang San, Battle of · 193
Matsushiro, Matsunosuke · 37
Mauchly, John · 224
Maurice Farman aircraft · 76
Maxwell, James Clerk · 13
Meitner, Lise · 143
Menzies, Prime Minister Robert · 130, 214
Mesopotamia · 30, 61, 64, 65, 66, 81
Metcher, Orm · 57, 61, 62
Meulstee, Louis · 149
MI6 · 140
Michelson, Albert · 13
Microlight radio · 285, 289, 316
Middle East campaign, WW1 · 65-9
Middle East campaign, WW2 · 128-9
military networks · 209, 260
miniature valves · 163
Minnetonka, SS · 35
Monash, General Sir John · 79
Montgomery, General Bernard · 129
Morse code · 12, 16, 17, 18, 21, 22, 31, 39, 84, 117, 136, 165, 167, 174, 182, 201, 202, 219, 221, 226, 227, 229, 234
Morse, Samuel· 20, 226
motor lorry-based wireless · 70
multiple tuner, 1904 · 83
Mustang P-51D · 193
N
Nagasaki · 132, 189
National HRO radio · 123
National Research Laboratory · 208
National Security Agency, NSA · 282
Nauen Wireless Station, Germany · 52, 140
Nelson, Lord · 36
networks · 226
Neupommern Island · 55
neutralisation · 162
Newfoundland · 16, 18, 40, 88
Nile, River · 65
Nipkow scanning disk · 110
Nipkow, Paul · 110
Nixon, General Sir John · 66
Norforce · 118
Norsk-Hydro at Vemork · 145-7
North Korea · 250
Norway · 146, 147, 150
Norwegian Resistance Museum· 147
O
O'Toole, Ian · 8
off-the-shelf · 271, 280
Operation Sealion (*Unternehmen Seelöwe*) · 77
Operation Grouse · 146-7, 149, 151
Operation Gunnerside · 147-8, 151, 152
Operation Jaywick · 155-6
Operation Swallow · 147
Oppenheimer, J Robert · 132
P
packhorse · 58, 95
Palestine · 61, 63, 64, 65, 66, 81
Panzer tank · 119, 126
Paraset replicas · 8, 175-88
Paraset (Whaddon Mark VII) · 141-2, 233, 310
Paris Peace Accords · 214
Passchendaele · 73
peacekeeping · 252
Pearcey, Trevor · 224
Pearl Harbor · 108, 130, 156
Penarth, Wales · 18
Percival, General Arthur · 130
Perikon detector · 73, 90, 105
Persia · 61, 64, 65-6, 67, 68, 81
personal computer, PC · 226
Phillips · 313
'phony war'· 126
Phuoc Tuy Province · 215
Pickard, Greenleaf Whittier · 89-90
Pickard, Major Roger· 7
Pintail radio · 238, 239, 277, 316
Plessey Corporation · 275, 276, 278, 280, 315, 317
point-contact transistor · 223
Poland · 36, 119, 120, 125, 126
Poldhu, Cornwall · 26, 28
Poldhu spark transmitter · 27
Polish Cipher Bureau · 121
Popov, Alexander · 17
Potsdam Conference · 190
Poulsson, Jens · 147, 151
Powerlite 110 · 264
PRC-F1 · 220-1, 234-5, 242, 273, 275
Preece, William · 14, 16-17
Pretoria · 32
'Prick 25' · 267
Princip, Gavrilo · 49
Project Parakeet · 276-7
Project Raven · 274
PRR-9 and PRT-4 · 220
public radio broadcasting · 109
Q
QRP transceiver · 242, 243
quenched spark · 75
R
Radio Australia · 185
Radio Corporation of America, RCA · 111, 267, 311, 313, 315, 317
Randall, John · 113
Raytheon · 281, 283, 284, 285, 289-90, 317
RCS Radio · 246
Receiver AR8 with transmitter AT5 · 156
Republic of Korea, ROK · 191
Rhodes, Cecil · 29
Ridgway, General Matthew · 191-2, 193
Rjukan, Norway · 146, 153
Rjukan Raid· 146-52
Roberts, Field Marshal Fredrick · 31

Index

Rogers · 313
Romani, Battle of · 65
Rommel, General Erwin · 128, 129
Ronneberg, Joachim · 147
Röntgen, Wilhelm · 12
Roosevelt, President Franklin D · 144, 146
Rorke's Drift · 288
Royal Airforce, RAF · 112, 113, 114
Royal Australian Air Force, RAAF · 131, 193, 215
Royal Australian Corps of Signals · 57, 115, 117
Royal Australian Engineers, RAE · 59
Royal Australian Regiment, RAR · 215-6
Royal Asutralian Signals Museum · 7
Rozhestvensky, Admiral Zinovy · 36
RT-F100 · 275, 276
RT-F200 · 275, 277
Rudram, David · 7
Ruhmkorff coil · 19, 21-2, 26
Rumania · 38
Russia · 34, 35-6, 250
Russian Revolution · 69, 108
Russo-Japanese War · 35-6
Rutherford, Ernest · 87
R valve · 73, 160

S

Saigon, fall of · 215
Salisbury Plain demonstration · 14-15, 20
Samsung · 265
Santayana, George · 290
Special Air Service, SAS · 219, 231
Scharnhorst, SMS · 142
Scherbius, Arthur · 120
Schonland, Sir Basil · 69
Sedd-al-Bahr Fortress · 60
SELEX/Marconi · 279, 291
self-loading rifle, SLR · 280
Seoul, bzattles of · 191
Serbia · 38, 45
Serong, Brigadier Ted · 215
ship's wireless of 1912 · 40
Shockley, William · 223-4
short magazine Lee Enfield rifle, SMLE · 281
Short-wave Receiver Mark III · 73

Short-wave Tuner Mark III · 73
Siemens & Halske · 14, 31, 32
Siemens coherer receiver · 31
Silesia · 107
Silicon Valley · 226
Sinai · 61, 65, 66
Singapore, fall of · 129, 130
single sideband, SSB, radio · 219, 220-1, 244, 270, 273, 274, 275, 282, 315, 317
Sino-Japanese War · 35
Slaby, Professor Adolf · 14, 29
Slaby-Arco · 14
software defined radio, SDR · 261, 268-71, 281-2
solid-state double-sideband transceiver project · 242-9
Somme, First Battle of the · 70
Somme, Second Battle of the · 79
South Africa · 20
South East Asian Treaty Organization, SEATO · 206
Southeast Asia · 140, 206, 210, 212, 214, 280
South Korea · 250
Soviet Union · 206
Spanish Civil War · 126-7
spark interrupter · 100-01
Sparrow Force · 156-9
Special Operations Executive, SOE · 141, 150, 311, 313
Sputnik 1 · 205-6, 211, 226
Standard Telephones & Cables, STC · 167, 313, 315
Steepholm Island, UK · 18
Steinmetz, Carl · 83
Sten gun · 148
Sterling transmitter · 75, 84, 95, 308
Storhaug, Hans · 147
Strassman, Fritz · 143
Stromsheim, Berger · 147
Suez Canal · 65
Sun Tzu · 257
superheterodyne · 92-4, 161
Sydney, HMAS · 54, 252
Sydney Progressive Amateur Radio Club · 242
Syngman Rhee · 190

T

Taiwan · 190
Taliban · 251, 254, 281
Tasma · 313, 315
TCP/IP · 209
TDMA · 284

Telefunken · 10, 14, 38, 70, 75, 84
Teleradio · 154
television · 110
Templer, General Sir Gerald · 198
Tesla, Nikola · 17
Texas instruments · 225
Thales Corporation · 277, 281, 317
theory of relativity · 13
Tigris River · 66
Timor · 156
Titanic, RMS · 38, 39, 40-1
Tobruk, seige of · 128
Togo, Admiral · 36
Togoland · 52
Tokyo Rose · 124
Tomlinson, Ray · 208
Toshiba computer · 264
Townshend, Major General Charles · 66
Trafalgar, Battle of · 36
transceiver · 163-4, 173, 234
transistors · 223, 229-41
Trans-Siberian Railway · 35
Transvaal · 29, 32
Travelling Caterpillar Fort · 45-9, 78
Triple Alliance · 50
Triple Entente · 49, 50
troposphere · 283
Trotsenburg, CK van · 32
Truman, President Harry S · 192
Tsushima, Battle of · 36
Tungtvannet, Kampen om ('The Fight for Heavy Water') · 147
Turing, Alan · 224, 265
Turkey · 38, 56
Type 3 Mark 1 · 142, 147, 149, 151
Type 3 Mark 2 · 141, 142
Type A Mark 2 'Ack' set · 116
Type A Mark 3 · 117, 118, 142, 143, 173, 174-5
Type C Mark 2 'Cork' set: · 117
Type TT-4 teletype · 263

U

Ukraine · 108
ultra high frequency, UHF · 138, 259, 261, 274, 281-2, 284
UNAMET · 253
undersea telephone cables · 52
unique resource locator, URL · 209

United Nations Security Council
· 190
United States of America, USA
· 50, 123, 132, 194, 206,
206, 312, 316
Unternehmen Seelöwe · 77
Urquhart, Major-General Roy
· 290
Uruzgan Province · 254
V
V-2 rocket · 205
valve · 73, 88–9, 90, 160–1,
169, 224, 229, 296
valved wireless in aircraft · 77
variable capacitance, VXO · 243
variable frequency oscillator,
VFO · 199
Vemork Hydro-Electric Power
Plant · 145, 146, 147, 148,
151, 153
Versailles, Treaty of · 107, 119
very high frequency, VHF · 138,
163, 172, 173, 194, 195,
199–201, 202, 216, 217,
219, 235–6, 237, 238, 239,
259, 261, 263, 272, 275,
276, 277, 278, 281–2, 283,
290, 313, 315, 317
Viet Cong · 213
Vietnam · 229
Vietnam War · 206, 212–21,
237, 259, 263, 280
Villa Griffone · 12
Vimy Ridge · 73
Vinton, Cerf · 209
Vrijstaat, Oranje · 29
W
Wagtail Project · 276
Wagtail Radio · 280
Walkie Talkie (see BC-1000
Walkie Talkie)
Wehrmacht · 119
Wells-Coates · 110
Western Front · 70
Westralia, HMAS · 252
Whaddon Mark VII · 141, 175
Wilson transmitter · 72, 73
Wilson transmitter · 8, 95–106
Wimereux · 18
Windhoek wireless station · 52
Winnie the War Winner · 156–9
Winterbotham, Fredrick · 121
Wireless Set C42 · 201, 202
Wireless Set No. 11 (Aust.) ·
134
Wireless Set No. 19 (Aust.) ·
169-70, 1951
Woolwich Arsenal · 73–4
World Heritage site · 84
Wozniak, Steve · 226
WS No. 1 · 117–8, 133
WS No. 9 · 133, 310
WS No. 11 · 117, 133, 134,
135, 136
WS No. 19 · 135, 169, 171,
194, 297, 310
WS No. 22 · 136, 162, 170,
173
WS No. 22 (Aust.) · 136
WS No. 62 · 139, 162, 170–1,
194
WS No. 88 VHF · 197
WS No. 101 · 117, 118, 133,
136, 157, 159, 163, 166–7,
239, 310
WS No. 108 · 128
WS No. 109 · 128, 133, 157,
158, 167–8
WS No. 122 Mark 2 · 135, 136,
137, 139, 162
WS No. 128 · 137–8
WS No. 133 · 134
Wynfawr wireless station · 85
X
X-rays · 12
Y
Yalu River · 191, 192
Z
Zedong, Mao · 190
Zeppelin airship · 112–3
Zimmerman Telegram · 92
Zworykin, Vladimir · 111